Constructive Assessment in Mathematics

Practical Steps for Classroom Teachers

David Clarke

KEY RESOURCES IN PROFESSIONAL DEVELOPMENT

Constructive Assessment in Mathematics: Practical Steps for Classroom Teachers

EDITORS: John Bergez and Steve Rasmussen
EDITORIAL ASSISTANTS: Romy Snyder and Sweigh Spilkin
COPY EDITOR: Greer Lleuad
PRODUCTION MANAGER: Luis Shein
INTERIOR DESIGN: Beverly Butterfield
COVER DESIGN: Susan Parini
TECHNICAL ARTIST: Jason Luz
CARTOONS: Ryan Alexiev
RESEARCH ASSISTANCE: Cathy Kessel

PUBLISHER
Steve Rasmussen

EDITORIAL DIRECTOR
John Bergez

10 9 8 7 6 5   00 99 98

Key Curriculum Press
P.O. Box 2304
Berkeley, California 94702
E-mail: editorial@keypress.com
Visit our Web site at http://www.keypress.com

Printed in the United States of America   ISBN 1-55953-201-7

PREFACE

Assessment in school mathematics is undergoing something resembling a revolution. In part new thinking about assessment stems from changes in the nature of school mathematics. But new thinking also arises because the nature of assessment itself is being questioned.

The National Council of Teachers of Mathematics *Curriculum and Evaluation Standards for School Mathematics* defines assessment as "the process of gathering evidence about a student's knowledge of, ability to use, and disposition toward mathematics and of making inferences from that evidence for a variety of purposes" (NCTM 1995, 3). This succinct statement leads to a number of philosophical and practical issues for teachers. At the practical level, what are some realistic strategies for gathering evidence about students' *knowledge of, ability to use, and disposition toward* mathematics? This book offers a sampling of such strategies and discusses some of the types of inferences that teachers can legitimately draw from the evidence they gather.

At a more philosophical level, the issue of purpose is central to our assessment efforts. I have chosen to use the term *constructive assessment* in part to highlight the question of purpose. Simply put, assessment should be constructive by being *helpful.* But assessment should also be "constructive" in the sense that it invites students to display their mathematical understandings through constructed rather than remembered responses. There are many examples of this distinction in this book; perhaps my favorite is the comparison of the two tasks "Name this shape" and "Write down five things you know about this shape." The information the teacher gains from the second task is much more likely than the first to offer insight into a student's understanding.

Two further considerations guided the development of this book. The first is that teachers' professional reading should be focused and purposeful. I have therefore aimed to present state-of-the-art classroom assessment as succinctly and as usefully as possible. The second consideration is practicality. All the strategies advocated in these pages have been used successfully by hundreds of teachers, within the constraints of a busy teaching schedule.

One last and very important point: We educators tend to view assessment from the teacher's perspective, but we also need to ask ourselves what purpose assessment serves for our students. This book attempts to address this question. After all, promoting students' learning is the central goal of our curriculum, and it is ultimately the students' actions that need to be informed by our assessment efforts.

Put all this together and you've got *constructive assessment.* As ambitious as it sounds, it's not that hard to do. If I can be permitted a word of advice, start small and go slow. There are many interesting and practical steps we can take to improve our assessment of students' learning of mathematics. Quite a few of them are outlined in this book. If things go well, you and your students will gradually come to engage in an increasingly rich exchange of assessment information.

David Clarke

ACKNOWLEDGMENTS

The contents of this book are the result of many conversations with colleagues and classroom teachers in several countries, to whom I extend my sincere thanks.

I would also like to extend my thanks to all the students I taught over my eight years of high school teaching. My commitment to practicality comes from the fading (but fond) memories of those classrooms.

The occasional good idea can usually be traced back to a conversation with my two children, Krisha and Nathan, who have provided important insights into the real issues of contemporary assessment in mathematics. Their tolerance is much appreciated.

In particular, I would like to thank John Bergez, whose sensitive and creative editing have given this book the major part of whatever coherence and accessibility it possesses.

D. C.

CONTENTS

INTRODUCTION

GUIDING PRINCIPLES OF CONSTRUCTIVE ASSESSMENT

Assessment occupies a central place in the mathematics curriculum. When assessment is done well, it can empower everyone, informing teachers about how to teach more effectively; informing students about what they have learned, what they have still to learn, and how best to learn it; and informing parents about how best to support their child's learning. Done poorly, however, assessment can misrepresent mathematics, misrepresent our students, and misrepresent our goals. At best, bad assessment may be simply uninformative, telling us little that improves our teaching and telling our students little that promotes their learning. At worst, it can be positively destructive, rewarding effort with failure and doing lasting damage to a student's confidence in his or her ability to understand and to use mathematics.

The idea that assessment can and should contribute constructively to the curriculum is a relatively new one. To realize the positive potential of assessment in our classrooms, we need a clear idea of why we are doing assessment in the first place, what it is we are assessing, and how best to go about it. Once we are clear about the *why, what,* and *how* of assessment, we can move on to the essential step of integrating assessment into our curriculum and our teaching seamlessly, as a central part of our daily routine.

This book is intended to assist teachers to find their own answers to the questions of assessment in mathematics and to begin to take practical steps toward making their assessment practices both constructive and effective. But what exactly is meant by *constructive* assessment?

Our answer to this question must take into account both parties in the assessment transaction.

For a teacher, assessment is a process in which we gather evidence, make inferences, draw conclusions, and act on those conclusions. Assessment is constructive when the focus of each stage of this process is the student's mathematical learning. In short, assessment is constructive when it assists us in fostering student learning.

For a student, assessment is an opportunity to demonstrate mathematical understandings and skills. More than that, it is a conversation with the teacher about what has been learned and what is still unclear and about what helped and didn't help the student learn. Assessment is an opportunity for mutual feedback and is a source of suggestions for action. In short, from the student's perspective assessment becomes constructive when it values what the student can already do and helps the student learn what she or he has not yet mastered.

Locating the student's learning at the heart of the assessment process distinguishes constructive assessment. This is the first guiding principle of this book. In the pages that follow, some additional key themes will emerge.

- ***Assessment should represent our instructional goals and values.*** Our assessment should reflect our knowledge and beliefs about what quality mathematical activity looks like. It should provide a major link between our instruction and the student's learning. In addition, assessment should anticipate the uses of school mathematics that will empower our students for a lifetime.
- ***Assessment is the exchange of information.*** Assessment should facilitate the exchange of information between teacher and student, and between other members of the school community. This information will relate to good mathematical activity,

Constructive assessment dispels the idea that "good performance" means "reading the teacher's mind."

effective mathematics learning, and effective mathematics teaching. Our selection of assessment strategies should ensure the exchange of quality information and help us maintain a constructive dialogue with our students about their learning and our teaching.

- ***Assessment should optimize students' expression of their learning.*** Assessment tasks should maximize the opportunities for students to express the outcomes of their learning rather than constrain them solely to the mimicry of taught procedures.
- ***Assessment should have instructional value.*** Good assessment should be synonymous with good instruction. It should therefore be possible to justify the use of an assessment strategy on instructional grounds. Students should learn something from their participation in assessment activities, and we should learn about students from their participation in instructional activities. In many cases, these activities will be indistinguishable.
- ***Assessment should anticipate action.*** Assessment techniques should be developed, selected, and implemented with the purpose of informing someone's actions. After all, no one has ever gotten any taller just by being measured. As teachers we must ask ourselves, How will this assessment activity promote and inform action by me, by other teachers, by my students, and by parents or other members of the community?

Central to this agenda is the substitution of portrayal for measurement as the underlying metaphor of assessment. Schools can no longer pretend that a one-dimensional number or grade can adequately or usefully characterize a student's mathematical learning.

- *Adequately:* According to the NCTM *Assessment Standards,* "Assessment is the process of gathering evidence about a student's knowledge of, ability to use, and disposition toward mathematics . . . ." Such a multifaceted portrayal cannot be given by a single measure, any more than we can portray a person simply by reporting his or her height.

- *Usefully:* Assessment is a process that serves a variety of purposes. A number or a grade does not provide sufficient detail to inform our actions as teachers, the actions of our students' parents, or the actions of the students themselves.

Until the need to *portray learning* (rather than to simply measure a performance) is recognized universally, teachers will have to initiate constructive assessment by selecting assessment strategies knowledgeably and purposefully and by effectively communicating assessment information. This requires a commitment, not to more assessment but to better assessment. For some teachers, a commitment to constructive assessment will mean working independently within an unsupportive environment to ensure that their assessment contributes positively to their teaching and to their students' learning. In doing so they will provide a model to other teachers of what is possible.

The process of implementing constructive assessment begins with our being clear about what we are trying to achieve. The remainder of this book elaborates on the general goals of fostering students' learning and improving teaching in terms of the three main functions of constructive assessment: Part 1 deals with the issues involved in *modeling good practice* through our assessment; Part 2 sets out a range of practical strategies for *monitoring good practice;* Part 3 discusses how our assessment can *inform good practice.* Appendix A provides some references to support a teacher's dialogue with other members of the school community about innovative assessment. Appendix B provides a range of mathematical tasks with both instructional and assessment potential.

PART 1

MODELING GOOD MATHEMATICAL AND EDUCATIONAL PRACTICE

The first function of assessment is to model good practice. As applied to mathematics, the basic principle can be stated quite simply: *assessment should model the mathematical activity we value.* But it is not only mathematical practice that students learn from our chosen methods of assessment; they also learn about mathematical *learning* and indeed about educational practice in general. For this reason assessment should also model *good educational practice.* This section sets out the major aspects of the role of assessment in modeling good practice.

Modeling Good Mathematical Practice

It is to be hoped that we value all that we assess, but do we yet assess all that we value? There is a belief within the mathematics education community that many previous forms of assessment have misrepresented mathematics. For example:

> Assessment based solely on closed or multiple choice items will be insensitive to the process outcomes which constitute much of contemporary mathematics curricula (NCTM 1989).

Implicit in this statement is a call for new methods of assessment sensitive to process outcomes. Process outcomes are one component of the new vision of what it is to do school mathematics. This new vision encompasses the basic tool skills that dominated previous mathematics curricula. In addition, however, school mathematics now includes a number of additional skills:

- Applying mathematical tools and skills in familiar and unfamiliar contexts
- Selecting and synthesizing these skills to solve novel problems
- Participating effectively in collaborative groups undertaking extended problem solving or project work
- Preparing written reports of such problem solving
- Using mathematics to communicate ideas
- Reflecting regularly, systematically, and critically on one's learning of mathematics

- Using available technology discerningly and appropriately to complete mathematical tasks
- Deploying skills previously associated with research, such as problem posing, experimental design, data collection and analysis, and the evaluation of findings

Clearly there are many facets to what we now think of as mathematical activity. Yet the major part of this new curriculum has not been addressed in any recent assessment techniques. If we genuinely value this rich diversity of competencies, then we must give each the recognition of assessment. Only then will our assessment truly model the mathematical practices we value. In the discussion that follows, we focus on how assessment can model the language of mathematics, the appropriate use of mathematical tools, and sophisticated mathematical activity.

Assessment should model the appropriate use of mathematical language

Mathematics is a language to be used purposefully, with meaning. At least three forms of mathematical language activity should be recognized and modeled by our assessment methods and strategies:

- *Reading mathematics:* Reading technical mathematical language with comprehension. This includes the interpretation of mathematical symbols, algebraic equations, mathematical diagrams, and graphs.
- *Interpreting situations mathematically:* Transforming everyday language into mathematical language. This includes representing everyday situations numerically, symbolically, diagrammatically, and graphically.
- *Expressing ideas mathematically:* Expressing mathematical understandings and ideas in mathematical language. This includes mathematical modeling, problem solving, demonstrations and proofs.

The responsibility of assessment to model these facets of the use of mathematical language can be seen in the following three task types.

- Write a story about the information shown in this graph.

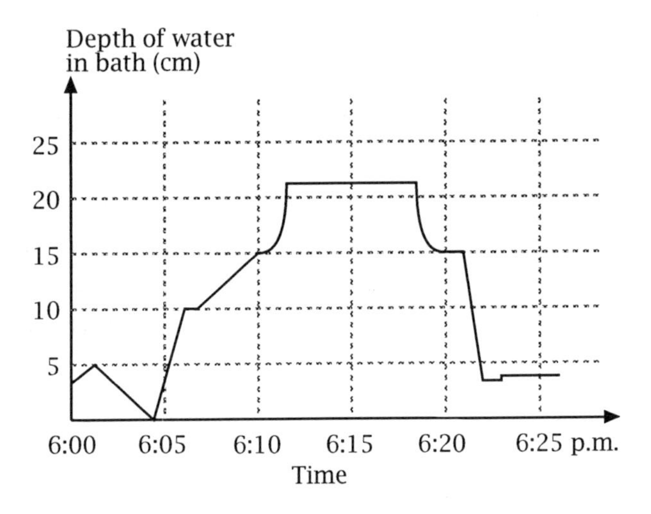

- Rewrite this article. Make use of tables and graphs where possible.

A Population Boom — In Cars

The world's population of cars is booming— and experts are worried about the consequences

In the middle of the 20th century, there were about 50 million cars and 2.6 billion people on Earth. By the mid-1990s, the human population more than doubled, but the number of cars increased more than 10 times, to 500 million. By 2020, the world's roads could be clogged with more than 1 billion cars.

The population boom in cars is especially noticeable in developing countries. In Thailand, for example, the car population shot up from 50,000 to 1.1 million between 1960 and 1990.

Other countries showed similar explosive growth in the number of cars over the same 30-year period: China, from 30,000 to 1.8 million; South Korea, from 40,000 to 2.1 million; India, from 300,000 to 2.3 million; Czechoslovakia, from 200,000 to 3.2 million; Mexico, from 600,000 to 6.8 million; and Brazil, from a mere 50,000 to a whopping 12.1 million.

The implications of the worldwide explosion in gas guzzlers are worrisome. Today, motor vehicles of all types account for almost half of all the oil used on Earth. Nearly one-fifth of greenhouse gas emissions comes from cars, trucks, and other vehicles. In almost half of the world's cities, exhaust from tailpipes is the single largest source of air pollution.

Adapted from "The Billion-Car Accident Waiting to Happen," *World Watch,* January/February, 1996.

- Write one paragraph explaining why this cartoon is supposed to be funny.

CALVIN AND HOBBES © 1988 Watterson. Dist. by UNIVERSAL PRESS SYNDICATE. Reprinted with permission.

Assessment should model the appropriate use of mathematical tools

Previous approaches to assessment have been concerned primarily with measuring the extent to which students possess a set of tools and occasionally the extent to which they can apply them. But to be mathematically equipped, a student must also understand the nature of mathematical tools and be able to select the correct tool for a given problem-solving situation.

A simple hierarchy of tool use can help us structure a fuller assessment of students' ability to use tools appropriately. The following discussion takes "finding the average" as an example of such a mathematical tool.

- *Tool possession.* Our assessment can tell us whether a tool is in the student's toolbox.

  Find the average of 19, 21, 23, 25, and 27.

- *Tool understanding.* Our assessment can suggest whether the use of the tools is well understood as a procedure

  The average of five numbers is 23. What might the numbers be?

  or even as a concept:

Using only the paper streamer provided, find and display the average height of your group.

- *Tool application.* Our assessment might even suggest whether the skill can be applied in a real-world context.

  Janet is the sales manager of a department store. She must maintain average daily sales of at least $8,500. Sales for the first four days of the week are $7,530; $8,475; $6,550; and $7,155. The store is not open on Sunday. What sales will Janet need to make on Friday and Saturday combined to come in over the target? Discuss whether it is likely that Janet will achieve her target.

- *Tool selection.* A toolbox is of little value if the student never chooses to open it. Our assessments need to be designed to tell us what mathematical tools the student chooses to use. Tool selection tasks must be amenable to solution through the use of particular mathematical tools, without specifying the tool required.

  It is Andrea's thirteenth birthday today. How many other children in the United States have their thirteenth birthday today? Show your method of solution clearly and identify any assumptions you made in solving the problem.

  Students' selection of the "average" tool should be evident in such calculations as the average number of birthdays per year.

Assessment should model sophisticated mathematical activity

One hundred years ago, twelfth grade examinations included such questions as "Add together 2.75, 8.5, and 1.125," "Extract the square root of 15,129," and "Subtract $2a + b$ from $4a + 3b$."

Where these activities are still considered relevant, we would find them today in the eighth grade syllabus. The mathematical performances required of our twelfth grade students today are much more sophisticated. But *sophisticated* does not just mean "more difficult." It should be possible for a student at any grade level to engage in mathematical activity that we would consider sophisticated or high-quality mathematical activity for a student at that grade level. We must be clear in our teaching what it is that we consider to be sophisticated mathematical activity, and we must provide through our instruction and assessment activities the opportunity for our students to engage in such activity.

Assessment should model not only the objects and ideas of mathematics, but also the relationships between these ideas. Our principal goal is to help our students develop their capacity for sophisticated mathematical thinking. It is essential that we model for our students just what sophisticated mathematical thinking looks like in our classroom and that our particular community of scholars shares a common understanding of what it is to think mathematically (whether in the third or the tenth grade).

Sophisticated mathematical thinking has several facets and can be thought of in several different ways. Our choice of assessment task will depend upon the type of sophisticated mathematics we wish to monitor and to model. In the following examples, sophistication in the use of graphing as a mathematical technique is explored in different ways.

- *Abstraction.* Sophisticated mathematics can involve the capacity to disengage a mathematical entity from a situation, recognizing the underlying mathematical idea.

  By visiting your school's library, your city library, or by contacting any appropriate people, gather information on the population of your school and your school district for at least the last five years. Analyze the data and in a paper show your careful mathematical analysis of the changes in your school's population. Then, based on your analysis, make predictions about your school's population growth over the next twenty years. Offer suggestions to your school board about how to deal with any problems your predictions (if true) might forecast.

- *Contextualization.* One measure of sophisticated mathematics is the student's familiarity with a mathematical object or procedure in many diverse contexts, since this familiarity will facilitate the solution of nonroutine problems and will optimize the transfer and application of the mathematics. The following examples of tasks focus on proportional reasoning.

Task 1: It takes 6 kg of pumpkins to make 4 pumpkin pies. How many kilograms of pumpkins would you need to make 10 pies? To make 135 pies? Use a graph to represent the amount of pumpkins needed to make up to 200 pies.

Task 2: A photograph is 4 inches wide by 6 inches long. If an enlargement of this photo is 10 inches wide, how long is it? If the enlargement needed to fill a billboard is 8 feet wide, how long would it be? Use a graph to represent the photo's length for any width up to 12 feet.

Task 3: The biological specimen Geomeuricus sequencius is born 5 cm in length. On the second day it grows 3 cm. The third day it grows 1.8 cm. On each following day it grows 60% of the previous day's growth. What is the length of the critter after two weeks? What is the limit on the length it could grow? Represent your answers by using a graph.

- *Interconnectedness.* A student's mathematical sophistication can also be measured by the interconnectedness of the student's mathematical knowledge, by the ability to change representations and to recognize similarities. Consider the following tasks that focus on linear equations.

Task 1: A recipe book suggests that the time required to cook a chicken is 20 minutes plus 15 minutes for each pound of the chicken's weight. Represent this rule as a table, as a graph, and as an equation.

Task 2: A recent article by Sonya Ross of the Associated Press states, "The world's big cities are growing by a million people a week and will hold more than half the Earth's population within a decade, the World Bank said Monday." Find a way to represent this information graphically so as to stress the significance of the statement. Represent the same information in the form of a table and then as an equation. Discuss the relative merits of the three forms of representation.

Modeling Good Educational Practice

In addition to modeling the mathematical activity we value, our assessment can and should model good educational practice. Whatever lessons we *intend* to convey about what is important in mathematics learning and education generally, our assessment deeds will speak louder than our words. As teachers we need to be sensitive to the messages—implicit and explicit—we are conveying about the purpose of education and the goals of schooling by the ways in which we choose to assess students' work. Let's consider several ways in which assessment can model the educational practice we value.

Assessment should distinguish between teaching and learning

We know that in a class of thirty individuals, each will emerge from a lesson with a different understanding. It is this learning that we want to reveal through our assessment. We know what we *taught*. What we do not know is what was *learned* by each individual. All we can be sure of is that the learnings of a class of thirty students will be different from one another and different from the learning that we shared through our teaching. Effective assessment reveals the differences between what was taught and what was learned.

Many teachers recognize, for example, that each individual constructs a personal meaning for the mathematics encountered in the classroom. For example, while *average* means "middle" to one student, it may mean "most common" to another. Many of these meanings will have good mathematical foundations, and the class will benefit from discussing them. Our assessment can reveal these differences in meaning. In doing this, our assessment is modeling an important educational distinction. Both teachers and students need to recognize that learning, even within a single topic, can take place in different forms, many of which will be mathematically valid.

In the past, assessment has offered a student two options: Replicate your teacher's understanding or don't respond at all. If we use tasks like "Name this shape," insight is likely to occur seldom and by accident rather than by design.

Insight by Accident

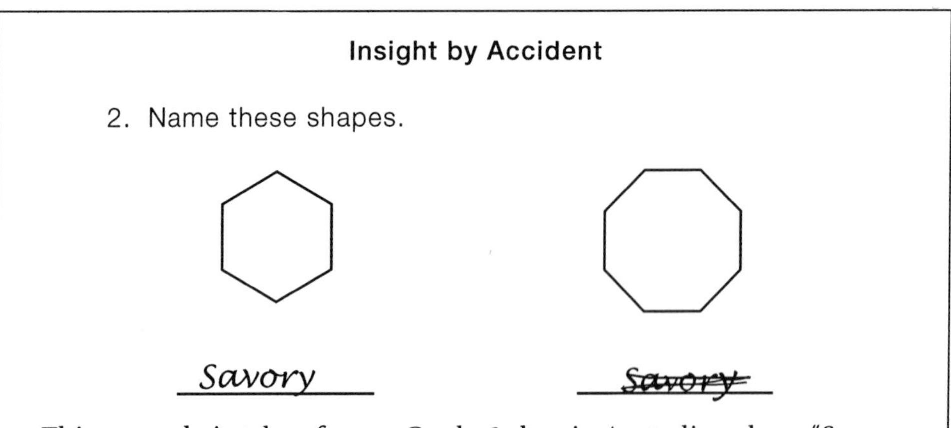

This example is taken from a Grade 6 class in Australia, where "Savory Shape" is the name of a salty cracker. What do you suppose was going through the mind of this student during two weeks of instruction in naming shapes? If you were the teacher of this student, how would you respond?

However, with only a minor change in the format of the task, we are much more likely to access the student's learning and less likely to restrict the student to imperfectly replicating our teaching. In the process, we will be conveying a respect for diverse forms of learning.

Insight by Design

Write down at least five things you know about this shape.

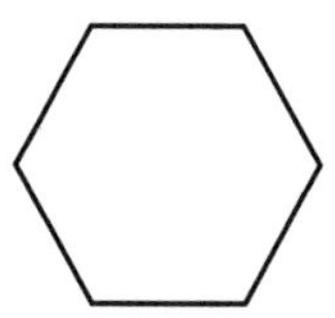

The shape's name may be one of the things to which the student attaches meaning, but there may be other properties of the shape with which the student is familiar, even though these were not dealt with in class. It is the student's learning that is of interest, both to the teacher and to the student.

Assessment is the exchange of information

When we pose a task in the classroom, we initiate a conversation. In effect we are saying not only, "What do you know about this matter?" but also, "This is important mathematics. This is what I value and hope you will value." The student's response continues the conversation: "Well, this is what I know, and I value the activity sufficiently to give it this much care." The conversation does not stop there. In this exchange we have both social and professional obligations. Socially we must acknowledge the answer in a manner likely to encourage the student to continue to contribute. Professionally we have an obligation either to provide feedback ourselves or to provide the opportunity for members of the class to join the conversation.

Our response to students' contributions is a powerful determinant of the way they come to view school, mathematics, and themselves as learners. Consider a conversation in an eighth grade science classroom:

Teacher: What is it that snakes don't have?

Student 1: Legs.

Teacher: No. That's not what you were taught. Who can remember what you were taught?

Student 2: Eyelids.

Teacher: Good for you! Yes, the answer is eyelids.

Neither of the students in this conversation may remember later on that snakes do not have eyelids. Both students, however, are likely to assimilate the very clear message that the purpose of a teacher's questioning is not to discover what they know, but whether or not they can find, within what they know, the magical "right answer"—that is, the one they were taught. This is not the purpose that our questioning should convey.

The original question has a whimsical richness that could have led to a fascinating exploration of what the class really did know about snakes. The opportunity was there for an exchange of rich and valuable information: telling the teacher what the students really knew and telling the students that it was their learning that was central to the conversation.

Assessment establishes the terms of the Didactic Contract

Every classroom functions on a set of shared understandings. Through these understandings the teacher and the class know what is expected of them and what their obligations are to the other participants in the classroom: what quality of work and effort teachers can expect from students, what assistance the students can reasonably expect from the teacher, what assistance the students can seek from one another, what level of explanation the teacher is obliged to provide, what questions the teacher can reasonably ask, what form of response will be considered satisfactory.

As Brousseau said, "This system of reciprocal obligations resembles a contract" (1986, 51). Students and teachers spend significant amounts of class time establishing the terms of what Brousseau has called their Didactic Contract. The terms of this contract are seldom expressed explicitly. They are inferred by all the participants in the classroom from one another's actions. Examining the implicit terms of the Didactic Contract can reveal what goals and values we are modeling in our classroom practice.

Although nowadays it is sometimes used pejoratively, the word *didactic* has an honorable pedigree. It is derived from the Greek *didaktikos,* which means "apt at teaching, instructive." Significantly, in modern educational usage (particularly in Europe), *didactic* has come to signify a less one-sided vision of education. It now describes a reciprocal relationship between teacher and student that produces learning as the result of a two-way interaction. Thus, the term *Didactic Contract* refers to the understandings shared by the parties to the learning transaction about what constitutes learning and educational success.

One of the clearest expressions of the Didactic Contract in any classroom is the way the teacher assesses learning. Many of the new tasks finding their place in today's mathematics classrooms challenge the prevailing Didactic Contract to which students and teachers alike subscribe. I will argue later that changing assessment necessitates renegotiating the Didactic Contract and that this renegotiation must be *visible* and *explicit*.

Assessment should anticipate action

The commitment to inform consequent action is the essence of constructive assessment. Constructive assessment is assessment that contributes visibly and helpfully to learning and teaching. And when we engage in such assessment, we convey to students that what matters is not so much past success or failure as what can be learned from the past that will improve their learning in the future.

The commitment to constructive assessment requires a commitment to quality information and quality communication. Suppose we pose the following task:

> Write the equations of five lines passing through the point (2,1).

This task has the potential to offer significant insight into student understanding, but this insight will be lost if our documentation and feedback are restricted to a numerical score or a grade. Suppose a student produces five lines but does not include the lines $x = 2$ and $y = 1$. To record a score of 5/5 in this case is to fail to identify a possible blind spot in the student's understanding. Or suppose a student responds (and many do), "This task cannot be done because there is insufficient information." To record a zero against the student's name does not inform either the student or anyone else—except to communicate failure. Moreover, the recorded score will tell the teacher very little at a later date that might offer insight into the student's learning or into ways that the teacher might improve her or his instruction.

Improving the quality of our tasks gains us little if our documentation and reporting methods discard the additional information such tasks provide.

A more helpful form of documentation will record what each student can and cannot do, and then communicate this information to the student in a way that informs subsequent learning. Parts 2 and 3 of this book outline practical steps toward achieving this result.

Assessment is something teachers are doing all the time

Historically assessment has been seen as a separate activity from instruction, and only forms of activity that involved written student product under testlike conditions have been sanctified with the label "Assessment." An even more restricted view of assessment tends to associate assessment only with activities that lead to the grading of student performance. In short, we have tended to label as "assessment" only a certain subset of our information gathering and exchanging. Is it any wonder that so many students have come to prize earning a grade over genuine learning—or even to confuse the two?

In reality, because assessment is the exchange of information in relation to a student's learning,

- a teacher observing a student at work is assessing,
- a teacher engaging in a class discussion is assessing,
- a teacher talking to a student about his or her performance is assessing.

For the purposes of this discussion, we will call these kinds of actions *observational assessment.* This usage involves interpreting *observation* in the broadest sense, to include, for example, purposeful classroom conversations with individual students. Such observational assessment is seldom associated with any sort of grading scheme. Yet the assessment information available to teachers through this type of classroom interaction is rich and insightful.

The concept of observational assessment highlights a point that has been implicit throughout this discussion: the widespread confusion between assessment and grading. Grading is not assessment; it is one—fairly simplistic—form of *coding assessment information.* As I have been suggesting, there are many other ways of representing and exchanging assessment information. Assessment should not be identified exclusively with a single method of representing assessment information, and the process of assessing learning should certainly not be confused with the activity of coding the results of our assessment. When this confusion occurs, it merely invites students to confuse learning achievement with achieving a good score. Compound this with the lesson that the way to a good score is to anticipate the desired response, and we are well on the road to instilling a view of education that many of us lament in our students, perhaps not realizing how much our practices have encouraged them to adopt it.

As teachers, we are assessing all the time; we can hardly refrain from assessing. It is encouraging to think that we may not need to introduce many additional assessment strategies, but that we need only to give due recognition to the activities in which we are already engaged. And when we consciously prize all the forms of assessment, we are much more likely to convey to students that what matters is what and how they are learning.

Assessment is a process.

Grading can be one product of that process.

The two should not be confused with each other.

All our different forms of assessment should be mutually enriching

Once we acknowledge the many forms of assessment, the next step is learning to use them in combination. The point is not that tests are "bad" and "alternative" assessments are "good," but rather that to draw conclusions about a student's learning on the basis of a single source of information is to run a high risk of misrepresenting that learning. Classroom observations and tests, to take two contrasting forms of assessment, both have their strengths and weaknesses. Used together, they offer both a richer body of information and a reciprocal validity check. Moreover, students whose learning is assessed in multiple, complementary ways will have a much rounder view of what learning is than those who associate educational success with good test performances.

The remainder of this book is intended to help teachers employ a broad range of assessment strategies both to improve the quality of information at their disposal and to provide safeguards against inappropriate judgments of student learning. Schools can institutionalize such an obligation by requiring that all students receive a written report that makes explicit reference to at least three different sources of assessment information. Strategies for integrating diverse forms of assessment information into a useful communication (such as a portfolio or a school report card) are discussed in Part 3.

Undertaking multiple forms of assessment does not have to mean a dramatic increase in the teacher's workload. Teachers who accept the responsibility to diversify their assessment practices may have little need for additional classroom activities. They may only need to recognize that the information exchanged each day about the success of the instruction and the development of the student mathematically is an important form of assessment. The decision is not so much whether to engage in additional assessment but how to publicly and systematically accord the same significance to *all* forms of classroom assessment that in the past has been accorded to more formal methods.

The principles and priorities outlined in the preceding pages can guide and structure our assessment activity in a way that acknowledges the modeling function of good assessment. But we do not live by principles alone—we also need specific, practical assessment techniques for putting principles into practice. As we turn to the other two functions of assessment—to monitor and to inform good practice—we will encounter several such practical techniques.

PART 2

MONITORING GOOD PRACTICE BY TEACHERS AND STUDENTS

The monitoring role of assessment is probably the most well-established of its functions. In a sense, Part 1 dealt with *what* assessment should monitor. Part 2 addresses the issue of *how* this monitoring might best be undertaken.

Choosing the Right Task

The Didactic Contract is characterized by reciprocal obligations between the teacher and the student. This aspect of the contract is most explicit in our assessment practices. I would like to suggest that a key clause in the contract is what might be called the Demonstration Clause:

> *It is the student's responsibility to demonstrate understanding and the teacher's responsibility to provide the opportunity and the means for that demonstration.*

Recognition and acceptance of these reciprocal responsibilities will focus both the teacher's and the student's efforts.

Constructive assessment must incorporate a sufficient range of tasks to meet our obligations under the Didactic Contract. In particular, such assessment must attend to language, tool use, level of sophistication, task type, context, and communicative mode. No single task can adequately address all these dimensions. However, an entire year's mathematics program structured to guarantee equitable representation of each of these dimensions in the selection of both instructional and assessment tasks would be unlikely to misrepresent either mathematics or the student.

Teachers choose assessment tasks on the basis of a set of criteria that vary from teacher to teacher, from class to class, and from topic to topic. The range of tasks that are now available is both challenging and exciting (see Appendix B). How, then, do we choose the right task for our purpose?

Different modes of assessment require different tasks. The wrong task will restrict, or even negate, the value of an assessment activity. For instance, a trivial procedural task offers little in the way of insight when used as part of an observational assessment. Conversely, an extended nonroutine problem can misrepresent the student's learning if administered as part of a time-restricted examination. We need criteria to guide the selection of suitable tasks that match our purpose and our chosen method of assessment.

One set of criteria relate to the effective modeling of quality mathematics. These criteria involve mathematical language, mathematical tools, and sophisticated mathematical thinking, as discussed in Part 1. Here we turn to criteria that include the type of mathematical performance, the diversity of task context, and the mode of communication.

Type of mathematical performance

Most teachers organize their lessons around mathematical content, and most curriculum packages take their structure from a sequencing of this content. (Some mathematics curricula depart from this model by adopting an integrated or thematic approach in which lessons are linked by a common theme or a common context rather than by a mathematical topic.) When the mathematics curriculum is organized into content-specific topics, such as "Graphing linear equations" or "Similarity and congruence," assessment is likewise structured to document appropriate performances within the content categories specified by the curriculum. Some novel problem-solving tasks, however, require the use of combinations of mathematical-tool skills drawn from different domains. Many argue that such tasks are useful precisely because they require students to demonstrate a capability to select appropriate tool skills and to combine them in a suitable solution process.

Scheduling such tasks in a curriculum organized by content topics can be difficult because the teacher cannot necessarily anticipate the mathematical skills or concepts that will be invoked. However, such problem-solving tasks represent a distinct task type and an important component of quality mathematics. As such, we should schedule them not according to the mathematical skills they require but according to the type of task and the type of mathematical performance required from the student.

The underlying principle here is that contemporary mathematics curricula should represent a model of valued mathematical activity. Our assessment must similarly constitute a valid and appropriate representation of school mathematics. To construct a mathematics program that includes assessment with representational validity, we must identify the dimensions across which representation must be maintained. Type of mathematical performance is one such dimension.

The discussion of the hierarchy of tool use in Part 1 can be used to distinguish types of mathematical performance. The simple instructional algorithm shown in the box Sampling Performance in the Spiral Curriculum can serve as a sampling frame for selecting tasks representing these performance types. Although our aim should be a balanced diet of tasks, this does not mean that every spoonful must be a balanced spoonful. For instance, it would clearly be impractical to assign an investigative project for every topic introduced over a year's mathematics program, but in this time students should have the opportunity to engage in each type of performance on several different occasions. We should consider, however, whether students should be required to engage in all types of performance in relation to the same mathematics content. It seems reasonable to suppose that a student's ability to use a particular mathematical skill in solving a nonroutine problem would depend to some extent on the breadth of experience the student has had in using the particular skill in a variety of situations. The notion of the spiral curriculum illustrated in the box assumes that during their school career students will revisit particular skills or concepts at increasing levels of sophistication.

The spiral-curriculum algorithm suggests a hierarchy of performance types corresponding to the four modes of tool use described in Part 1. It thus identifies four distinct forms of mathematical performance for assessment purposes. The hierarchy of performance types can also provide a practical structure for a weekly worksheet to be completed at home. (Of course, the worksheet might be administered less frequently than weekly; it is the structure that is important.)

Structure for a Weekly Worksheet

Part A	Four *routine tasks* from this week's content
Part B	Three *Good Questions* on related content from last year
Part C	Two *applications questions* requiring content from two years ago
Part D	One *challenging problem* linked to content from three years ago

This approach assumes that sophisticated performance requires ongoing experience in the use of a concept or a skill. Some students will develop the capability for sophisticated performance more rapidly than others, and more rapidly than is suggested in the algorithm. An assessment program should provide the *opportunity* for students to demonstrate each type of performance. Whether each type of performance is *required* is determined by the curriculum or the teacher's discretion.

Sampling Performance in the Spiral Curriculum

Students should be provided with the opportunity to do the following.

a. *Solve novel problems* requiring the mathematics encountered three years ago (tool selection)

A golf ball is struck so that it follows the path shown. Draw a graph to represent the ball's speed throughout the time it is in motion. Write a paragraph explaining the shape of your graph.

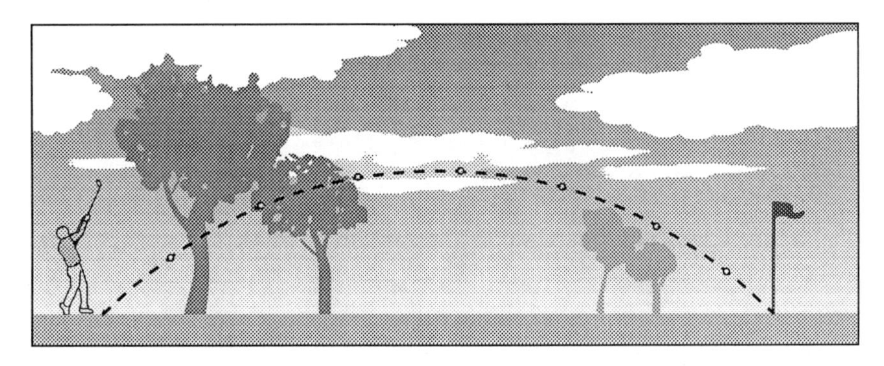

b. *Apply in familiar situations* the mathematics encountered two years ago (tool application)

Taxi fare is calculated by adding "flag fall" (a fixed cost for getting into the taxi) to a "travel cost" (cost per mile multiplied by the number of miles traveled). Anne's taxi fare was $24 for a journey of 10 miles. Suggest possible values for the flag fall and the travel cost, identify your preferred combination, and justify your choice.

c. *Demonstrate understanding* of the mathematics encountered last year (tool understanding)

A line passes through the point (2, 1). Find at least five possible equations for this line.

d. *Recall and replicate* the mathematics encountered this year (tool possession)

Find the point of intersection of the lines $y = 2x - 1$ and $y = 12 - x$.

One concern voiced by teachers is that there is not a sufficient range of tasks available to model these different types of mathematical performance. Appendix A provides sources that teachers can consult to enlarge their repertoire of mathematical tasks. Appendix B offers a Continuum of Mathematical Tasks spanning twelve task types with one example of each. But the most useful source of tasks is the classroom teacher. A teacher-generated task will match more accurately the needs and competencies of the class and consequently will provide more useful information to both teacher and student.

Some teachers lack confidence in developing open-ended tasks. One teacher remarked

> I need a guide to creating open-ended questions. Teachers (at least the ones I work with) do not know how to convert traditional-type questions into an open-ended format.

"Good Questions" represent one type of open-ended task that can easily be developed by adapting conventional questions (Sullivan and Clarke 1991). Good Questions are content-specific but are sufficiently open to allow most students the opportunity to display what they know across a range of levels of sophistication. Here are some examples of this type of task.

> Write the equations of at least five lines passing through the point (2,1).
>
> A number is rounded off to 5.8. What might this number be?
>
> The average of five numbers is 17.2. What might the numbers be?
>
> Draw a triangle with an area of 12 square units.

All these tasks have the characteristic of supplying students with something that looks like the answer to a conventional task and of asking them to suggest possibilities that might lead to that outcome. Students are then required to explore the related mathematical procedure from a different direction. Each of the examples also has the virtue of allowing students the option of providing a single "solution," multiple solutions, or a general solution. As a result, these tasks can be discerning assessment instruments, distinguishing between levels of student response. For instance, in relation to the question

A number is rounded off to 5.8. What might the number be?

the following levels of response are all possible and are all informative:

- No response or an incorrect response
- A single value: 5.81
- Several values: 5.77, 5.79, 5.81
- A systematic list: 5.75, 5.76, 5.77, 5.78, 5.79, 5.80, 5.81, 5.82, 5.83, 5.84
- A class of solutions (more or less precisely defined):

  The smallest it can be is 5.75 and the largest it can be is less than 5.85
  $\{x: 5.75 \leq x < 5.85\}$

Students need to be made aware of these different levels of response and their relative merits, even though only some of them perform at the most sophisticated level. They should also be led to see the value in a general response. Nonetheless, the student who gives only a single numerical answer can have it affirmed because it also meets the conditions of the task.

Diversity of task context

Current thinking in mathematics instruction calls for setting mathematical tasks in specific contexts. An adequate sampling of a student's behavior requires that we vary these contexts. This contextual variation also holds the promise of facilitating the student's subsequent use of mathematics in diverse contexts outside the classroom.

One difficulty in using contextualized tasks for the purpose of assessment is deciding how much weight to give to the contextual coherence of the student's answer and how much to the demonstration of mathematical tool skills. Consider for example, the following task.

> Fred's apartment has an area of 60 square meters. There are five rooms in Fred's apartment. Draw a possible plan of Fred's apartment. Label all rooms and show the dimensions (length and width) of each room.

Alan's response looked like this (Clarke and Helme 1994):

	2 m	2 m	2 m	2 m	2 m
6 m	Living room	Bedroom	Bedroom	Bathroom	Kitchen

As Alan's teacher, what do you do? Does the mathematical consistency of his solution outweigh the contextual inconsistency? Clearly, no one would design a real apartment in which you had to go through the bathroom and two bedrooms to go from the kitchen to the living room. If your purpose was to assess Alan's understanding of the concept of area, perhaps his answer is satisfactory. If, however, you expected a more sophisticated plan and, as a consequence, a less trivial partitioning of the 60 square meters of area, then you might decide to penalize Alan for the triviality of his solution or for its contextual inconsistency.

The difficulty is that Alan may recognize the practical inadequacy of his design. It may be that he just did not feel that practicality was a relevant criterion in a mathematics class. In fact, when asked about his solution, Alan said

> I was just thinking of the math involved. I guess that probably comes from the way you learn math, just doing calculations and learning calculations. But if I was planning an extension for the house . . . I wouldn't think in that kind of way (Clarke and Helme 1994).

Thus, Alan's response may not indicate an inadequacy to contextualize mathematics successfully, only a lack of recognition of any need to do so. His response conformed to his understanding of the Didactic Contract as it operates in mathematics classrooms. If we are to assess Alan's mathematical performance fairly, then we must first establish a shared understanding with Alan of the significance to be attached to the context of a problem and of the extent to which his solution must take that context into account. That is, we must establish the terms of a new Didactic Contract with our students. (A procedure for doing so is presented in Part 3.)

Research suggests that student interaction with task context is highly personal. Some students find the location of a task in any context to represent an additional level of difficulty, while other students find the use of familiar contexts an aid to completing the task. For this reason, *both* the following tasks should be employed in the assessment of a student's understanding of the concept of average:

> The average annual maximum temperatures of two U.S. cities is 65°. One of the cities had a range in average maximum monthly temperature of 60° over the whole year, while the other had a range of only 20°. Suggest two cities for which this might be true. Give possible average monthly maximum temperatures for each city for each month throughout the year.
>
> The average of five numbers is 17.2. What might the numbers be?

In short, the teacher's obligation is to use tasks that sample a range of contexts and a range of contextual detail.

Modes of communication

Thus far we have considered diversifying types of mathematics and types of contexts in which tasks are situated. It is also important to vary the modes of mathematical communication called for in our assessments. One teacher voiced a common difficulty:

> [One of my major problems is] how to efficiently evaluate students who can do well when the work is oral, but have a terrible time with reading and writing.

This remark points to the need to include multiple modes of communication in our assessment if we are to get an accurate picture of students' understanding. It also draws attention to the need to support students when a task requires a response in a communicative mode with which the student has yet to develop fluency or expertise. Where possible and appropriate, students should be offered the opportunity to display their mathematical understandings through physical representation, numerical calculation, geometric construction, written narrative and explanation, demonstration, dramatic performance, and symbolic notation. They can be asked to perform tasks graphically, orally, visually, and electronically (through the medium of computer software). By using tasks that cover a range of modes of expression, teachers can accommodate a student's preferred mode and preexisting language skills while still accepting a responsibility to help the student develop competence in all communicative modes. It is neither necessary nor appropriate to employ all modes of communication in response to any one task or content area, but teachers should be receptive when students suggest that another mode of communication would enable them to demonstrate their understanding more effectively.

Mathematical projects

One of the characteristics of the new mathematics curricula is the increasing use of extended mathematical problems and investigative project work. This trend represents a valuing by the mathematics education community of a new form of mathematical performance of sufficient significance to warrant separate treatment.

Because of the extended nature of project work, such tasks represent an opportunity for the student to be given significant control over the form of the response, both in terms of context and in terms of mode of communication. A selection of suitable projects for high school students might include the following.

- *Approximation and error.* Investigate the issue of approximation in scientific measurement and relate it to experimental error. Your report should give specific examples of real situations in which approximation and error are important and should also make clear the role mathematics plays in understanding and dealing with approximation and error. Your report should include a written component but can also include diagrams, models, demonstrations, or some other form of verbal or visual presentation.
- *Symmetry.* Identify a profession (for example, engineering, architecture, art, physics, or choreography). Describe at least three different situations involving symmetry that are likely to be relevant to someone working within that profession. Make clear the mathematics required for an understanding of the role of symmetry in those situations. Your report should include a

written component but can also include diagrams, models, demonstrations, or some other form of verbal or visual presentation.

- *Probability.* Identify three situations in which a lack of understanding of probability would be likely to have serious consequences for some person's life, profession, or survival. The situations should be realistic, and you should portray them realistically, either in written form or by scripted performance. The report should (a) make clear the relevance of probability to the situation, (b) outline the consequences of a lack of understanding of probability, and (c) describe in detail how probability might be used to the advantage of the individual. Your report should include a written component but can also include diagrams, models, demonstrations, or some other form of verbal or visual presentation. Include all relevant calculations in your report.

Projects can take many forms and serve many purposes—these three examples represent one possible type. All of them offer students the opportunity to integrate related mathematical concepts and skills and apply them within a context of their choosing.

Another type of project, suitable for younger students, requires that they engage in the mathematics associated with a specific realistic task. Here are some examples of this type of project.

- *Parking lots.* Redesign the school parking lot to improve its efficiency, capacity, and safety.
- *Soda cans.* Design an aluminum soda can that holds an appropriate volume and is attractive, easy to handle and store, and economical to manufacture.
- *Excursion.* Plan and estimate costs for an excursion for your grade to a local place of interest.
- *Game.* Design a board game, including the playing board and a clear set of rules.
- *Vacation.* Plan and develop a budget for a vacation trip for a group of four people to a location within the United States.

As with the other projects, include in the instructions "Your report should include a written component but can also include diagrams, models, demonstrations, or some other form of verbal or visual presentation." Most teachers would also be prepared to accept suggestions from students of alternative projects of the same type that the student would prefer to undertake. One teacher described this scenario:

> Kids sometimes bring in problems or make suggestions. The maintenance man has been good about bringing in problems. Once he needed to drain a pipe and wanted to know how big a container he needed for the liquid in the pipe. The class calculated it for him. Another school still uses coal and has a funny shaped coal bin. They wanted to find its volume, so the students calculated that.

Projects offer a change of pace from the usual classroom mathematics activity. This can be a source of satisfaction and enjoyment to both student and teacher, as reflected in this comment from a mathematics teacher:

> We've gotten excited about projects. They're displayed in the classroom, in the front of the school. One of the projects is making a game.

As the complexity of the task increases, so does the complexity of any associated grading of performance. While it is difficult to specify assessment criteria in detail, it seems reasonable to require that students should accomplish the following:

- Meet the specific terms of the project
- Employ mathematics appropriate to their grade level
- Relate the mathematics to the project task in a sensible and purposeful manner
- Define all significant terms required by their solution or describe all situations in a clear and succinct manner
- Produce a response that is internally consistent (both mathematically and contextually) and true to the context as they have described it

A question that can be usefully asked of all students undertaking project work is, Does your project make sense?

The question that must be asked of the mathematics curriculum as a whole is, Does this curriculum offer every student the opportunity to learn important and appropriate mathematics? The essential point in choosing the right task—whether it involves the routine recall of a procedure, a challenging nonroutine problem, or an extended investigative project—is, Is the performance required by this task consistent with the goals of the mathematics curriculum? One additional question must also be asked: Does the assessment offer every student the opportunity to display the mathematics he or she has learned? We must be able to answer these three questions in the affirmative.

Time-restricted tests

A final observation on choosing the right assessment task concerns the use of time-restricted tests. Some mathematical performances simply require more time than is usually available in a conventional classroom test. Putting these kinds of items on a time-restricted test serves no valid assessment purpose.

Indeed, the imposition of a time restriction on an assessment activity is very difficult to defend on any grounds other than convenience. It is true that an excessive test time can create organizational problems for the teacher and motivational and concentration problems for the student. However, an inadequate test time runs the serious risk of giving us inaccurate information about the student's understanding.

If, for instance, we want to assess a student's capability to carry out an extended mathematical investigation, then a time-restricted test is an inappropriate measure. If we want to check what is in the student's mathematical toolbox, then a time-restricted test may be convenient and efficient. Time-restricted tests are appropriate only for a limited range of purposes. As with any assessment technique, teachers need to match the method to the purpose and to ask whether the information provided by the time-restricted test will be valid and useful. The information is valid if it accurately reflects the student's competence in or understanding of some aspect of the mathematics curriculum. It is useful if the information has the potential to lead to constructive action by someone: teacher, student, or parent.

Not all performances are appropriately assessed by time-restricted tests.

Observational Assessment

Earlier I emphasized that teachers assess all the time in the course of instruction. An important example of assessment is the purposeful observation of students engaged in instructional activity. There are three things that matter for the effective use of observational assessment:

- Appropriate task selection
- An assessment-friendly classroom
- A succinct and effective method of recording any emergent insights arising from the observation of the students

A scenario for observational assessment might look like this:

> The class is divided into small groups. A task is provided, typically open-ended and requiring more than just the recall of a taught procedure. Each group works collaboratively on the task and produces a group report of their solution. This report can be verbal or written.

One common approach is to require each group to report briefly to the entire class, and then, in the light of comment and criticism from their peers, to refine their report for submission to the teacher in written form. While the students are engaged in working on the task, the teacher can move from group to group, making notes concerning student behaviors to be encouraged or challenged, strengths to be acknowledged, and weaknesses to be addressed.

Any classroom task addressing legitimate mathematical content can provide insight into a student's mathematical learning. We have already discussed the criteria for choosing the right task. For the purposes of observational assessment, extended tasks that encourage student collaboration can be particularly useful. In the scenario just outlined, for example, any of the four following tasks would be appropriate.

Using only the paper streamer provided, find and display the average height of your group.

If one vertex of a triangle is allowed to move around a circle, what is the locus of the orthocenter of the triangle? (Explore using The Geometer's Sketchpad.)

How many elephants are there in the United States?

If $P(x)$ is a polynomial in x of the form

$$P(X) = a_0 + a_1x + a_2x^2 + \ldots + a_{n-1}x^{n-1} + a_nx^n$$

and $Q(x)$ is another polynomial in x of the form

$$Q(X) = a_n + a_{n-1}x + a_{n-2}x^2 + \ldots + a_1x^{n-1} + a_0x^n,$$

what is the relationship between the roots of the two polynomials $P(x)$ and $Q(x)$?

An assessment-friendly classroom

In an *assessment-friendly* classroom, the students are given both *responsibility* for aspects of their assessment and the *opportunity* to make decisions about preferred modes of assessment, to comment upon their learning and the teaching they have received, and even to devise test items or project topics. To optimize the effectiveness of your informal assessment, you don't need a special set of guidelines of things to look for or a special set of questions with which to probe your students' understandings. You do need a wide and rich selection of instructional activities. And, fundamentally, you just need to move around the classroom while these activities are in progress and talk to the students, trying to be sensitive to what is happening. If the activities are interesting and educative, and the classroom atmosphere is both challenging and supportive, then the exchange of useful assessment information is highly likely.

There are teacher behaviors, however, that can stifle the exchange of useful information or generate poor or misleading information. In particular, inadequate "Wait Time" has been identified as a major source of misleading information and mistaken judgments.

The principal issues regarding Wait Time are easily summarized. When asking a question of either the whole class or a small group of students, there are three crucial moments when you should pause.

1. After posing the question, and before naming the student you would like to respond. "What is the area of a rectangle 9 cm by 7 cm? [Wait Time 1] Steve?"
2. After naming the student and before making any other statement or comment. Many of us can be very bad at this. Rather than embarrass a student, we are inclined to leap in with a suggestion or to quickly redirect the question to another student. Give the student time to respond. [Wait Time 2]
3. After the student has replied. Allow time for everyone to think about the response before you make any comment, even to ask another student to respond. [Wait Time 3]

Wait Time may seem relevant only in whole-class situations, but the research into Wait Time has revealed important insights into some of our less constructive practices. For example, every teacher forms a mental profile of the mathematical competence of each student. Consider two students, one of whom you have "labeled" as very capable (Shokria) and one who seems to be struggling (Maria). Who might expect the longer Wait Time when responding to a question? Research suggests that it is the more able student who receives the longer Wait Time. Why? Well, you don't really think Maria is going to answer.

Her past performance in class has not suggested that success is likely, so you quickly move the focus of attention to Shokria, who you are confident will answer correctly. Perhaps Shokria is also a little slow to respond. Well, you hang in there anyway—she is a capable student; she will get there. As a consequence, Maria, given the task of responding in an inadequate period of time, has fulfilled your expectations, while Shokria, with the benefit of your tolerant anticipation of a correct answer, finally produces the correct response, as you knew she would.

The uneven distribution of Wait Time has been consistently documented in the research literature. Unfortunately, the messages exchanged in such classroom questioning do more than confirm teacher judgments; they can establish and consolidate negative student self-esteem. And they can be wrong! This is an important message of the Wait Time research, and it applies more generally to any teacher-student interaction.

In the assessment-friendly classroom, students can be heard. They know they will be given adequate time to gather their thoughts and formulate an answer. And, as their teacher, you know you will get a better quality of information.

Of course, observational assessment is only one of the many tools at a teacher's disposal. Even in the best conditions, judgments derived from informal assessment are subject to error, just as judgments derived from tests are. As I emphasized at the outset, we need a commitment to multiple modes of assessment, in part to provide safeguards against incorrect professional judgments. We will shortly discuss one way to monitor the consistency of the messages received through various assessment channels.

The essential message here is that constructive assessment is a visible and integral component of classroom activity, one in which students are invited to play an active role. Far from being covert, observational assessment is a visible public valuing of mathematics and of student learning.

Teacher recording techniques

Many teachers make use of checklists to record students' acquisition of key skills or concepts. But while such checklists can be an effective cumulative record, they can be an unwieldy tool in the classroom. A popular method of keeping track of the insights offered by instructional activity is the *annotated classlist.* The goal here is to record only information that challenges or extends the teacher's understanding of the student: the capable student who experiences unexpected difficulty or shows lack of understanding; the student you had labeled as "struggling" who shows unexpected insight or understanding; a sudden insight into the source of a student's difficulty; the emergence of new behaviors or capabilities not seen before, such as leadership, perseverance, understanding of a new concept, or a concern with accuracy.

Such insights are recorded succinctly on a classlist beside the student's name. Brevity is essential:

Difficulties with average!!
Reluctant to participate
Chronic lack of confidence
Effective group leader
Excellent estimation skills
Used percentage well
Good use of diagrams/ratio/Pythagoras
Insisted on checking answer

Teachers report that most lessons will typically produce four or five such insights. After a week, the classlist might look like the one shown in the Sample Annotated Classlist (facing page).

Teachers using annotated classlists find ways to highlight observations requiring urgent action and ways to indicate that this action was taken. All entries are acted upon in some way, if only by commending the student.

Teachers have also reported that the annotated classlist serves to highlight "invisible" students. A quick scan of the last one or two weeks' classlists will reveal students against whose names nothing has been written. By helping us identify students who are getting less than a fair share of our attention, the annotated classlist becomes one practical step toward maintaining equity in our classrooms. This is assessment informing action. This is constructive assessment.

Teachers have employed many variations of this basic technique. Some teachers record their insights on adhesive labels, date them, and at the end of the day insert them into a class record book in which a double page is reserved for each student. Other teachers simply maintain a file of annotated classlists

Sample Annotated Classlist

Week beginning August 30	COMMENTS (Aberrations and insights)	ACTION REQUIRED	ACTION TAKEN
Bielecki, Barry	No concept of odd and even	*	
Carlton, Steve	Showed leadership in the group		
Chin, Hong			
Clemente, Ricardo			
Cook, Wendy			
Delamere, Gina	Thought 63 and 36 the same	*	✓
Gonzalez, Jorge	Really tried		
Grace, Nathan	Sequencing problems	*	
Hsia, Albert			
Luey, Constance	Spatial thinker		
McGraw, Joan	Recognized significance of a counter-example		
Medrano, Omar			
Moule, Julia			
Musial, Stan			
Nevarez, Pedro			
O'Connell, Deirdre			
Ogden, Kate			
Palmer, Jim	M.A.B. work needed - hundreds and tens		
Pignatano, Joe			
Plank, Edie			
Rawlins, Carlene			
Reeves, Deon			
Ruiz, Nina			
Stephens, Kaye			
Stephens, Maxine	Decimal place value a problem (division)	*	
Stone, Stephanie			
Williams, Ted	Good at routine tasks - thrown by challenging ones		

for ready reference prior to parent-teacher meetings or for the writing of school reports.

The value of such observational assessment can be argued on three grounds:

- The information it provides is of high quality, drawn from more complex performances and not distorted by the stress sometimes associated with tests.
- The assessment is gained efficiently, without the disruption of ceasing instruction to hold a separate "assessment event."
- The information is accessed in a situation in which constructive action is still possible, unlike tests (which typically occur at the end of a unit or term or semester or year, at the completion of a topic or a course, when instructional action is difficult, inappropriate, or impossible).

Such assessment usefully augments test-based and other forms of assessment in the interests of more effectively portraying students' learning. As I have emphasized, neither tests nor observational assessment—nor any other particular form of assessment—should be accorded absolute authority. Rather, we should check continually to see whether our different assessment strategies are providing us with consistent information. For example, the simple comparative technique shown in the Assessment Crosscheck Technique box can provide a useful cross-validation of our observational assessment and our testing.

Assessment Crosscheck Technique

Obtain a classlist and draw up three columns.

- In column 1, record your *predicted* test score for each student.
- In column 2, indicate your *confidence* that the predicted score is within ten percent of the actual score by writing either C (confident) or U (unsure).
- In column 3, record each student's *actual* test score.

Columns 1 and 2 must be completed *before administering the test.* After the test, interview any student for whom the discrepancy between predicted and actual scores is greater than ten percent and ask the student to complete some of the tasks on which she or he experienced unexpected difficulty or success.

The chart below depicts one such assessment crosscheck.

Assessment Crosscheck

Name	*Predicted score*	*Confidence*	*Actual score*
Student 1	65	C	70
Student 2	50	C	48
Student 3	85	U	82
Student 4	80	C	66
Student 5	70	U	83
Student 6	60	C	62

What do the results suggest to you? Which sets of scores would cause you the greatest concern? If you were the teacher of these students, what actions would you take?

Consider these two anecdotes, both true, in which it is shown that both our testing and the conclusions we draw from our classroom questioning may be in error.

Anecdote 1: "Something about the test"

A teacher who was experimenting with this approach told me, "You know, I almost got them all right." A short conversation revealed that there were only three students, and one girl in particular, for whom the predicted scores and the actual scores differed by more than ten percent. "I just thought she would do better on the test." Encouraged to sit down with the student and get her to complete some of her incorrect test items in a one-to-one situation led to a telephone call a few days later: "You know, she really does know that stuff. I guess it must have been something about the test."

Anecdote 2: The delayed reaction

An interview with a child in Grade 2 revealed an almost comic delayed reaction in response to any question. The child would pause, look around, look down, put his finger to the side of his nose, stare blankly into space for awhile, and then give a thoughtful and sophisticated answer. It did not matter whether the question was "What is 3 plus 5?" or "How many eighths are there in two and one-quarter?" There was always a pause of more than twenty seconds, with associated body language, and then an articulate, thoughtful, and typically correct answer. Wait Time in a typical classroom is seldom as much as five seconds. What is going to happen to this child? In fact, when the teacher was encouraged to sit down and talk to the child, the results were so different from the teacher's expectations that the child was immediately given much more sophisticated mathematics in class and has continued to perform successfully. The origin of the child's idiosyncratic answering style? One day, I happened to ask his father a question. He paused, he looked up, he looked down, he put his finger to the side of his nose, he paused some more, and then gave a well-considered, articulate answer. Funny things, families.

Both anecdotes illustrate the perils of restricting your sources of assessment information to one strategy. Our classroom observational assessments can be accurate and rich with information, and our tests can be useful measures of skill acquisition. But both can be in error, and a simple strategy like the assessment crosscheck can identify important discrepancies. This straightforward comparison of the information you are receiving from your different assessment strategies will improve the quality of your observational assessment, the quality of your tests, and the quality of your judgments regarding students' learning.

Suppose you identified a need to improve the quality of your testing; what might you do? A first step would be to review your testing strategies. As with any form of assessment, the key consideration is, Could this test misrepresent the child's learning? An additional consideration is, Could this test misrepresent the mathematics I value? There are various factors that might lead to such misrepresentation, such as the artificiality of time restrictions. We can improve our tests by deliberately addressing the issues related to each such factor. In doing so, we are giving priority to the validity or authenticity of the assessment.

Assessment Activities

In addition to observational assessment, we need a range of practical procedures that provide the information we need to monitor the mathematical activity we value. The assessment activities described in this section have all been used by many teachers with success. In selecting from this menu of activities, teachers need to ask not only, Will this approach give me the information I need? but also, Can I put this strategy into operation without generating an unmanageable increase in my workload?

Practical assessment

A key issue in contemporary assessment is the demand for consistency between instructional activity and assessment activity. In informal assessment situations this is not an issue: the same tasks serve both purposes. In formal assessment, however, it has been possible to use one type of task for instructional purposes and another for assessment. This makes little sense. The performances with which we initiate our students into mathematical activity should be the same type of performances by which their success will be judged. For example, if a topic involves practical activity and a test is one of our intended assessment strategies, then we should include practical performance items on the test.

Many of our lessons involve practical activity: drawing mathematical designs, maps, or house plans; constructing geometric figures; measuring; collecting and analyzing probability or statistical data; using computers; or conducting experiments. Our belief in the educational value of practical activity is well founded. Practical activity can

- increase motivation and engagement,
- increase task and concept accessibility,
- increase task authenticity,
- increase the likelihood of transfer to contexts outside the classroom.

Take measurement, for instance. Students learn more about measurement by actually engaging in measuring. Measurement is, of itself, a practical activity. To represent it in the classroom wholly through pencil-and-paper activity would be to misrepresent it. Further, the contexts in which we would hope our students would make use of the concepts and skills taught within measurement are predominantly practical contexts. It does not make sense to assess by a pencil-and-paper test skills or concepts that have been taught through practical activity—unless we can include a practical element in the test situation.

In practical tests, student performance is particularly visible, both to the teacher and to the student. This visibility gives such assessment situations a communicative power that can inform student or teacher action less ambiguously, more usefully, and immediately. Two scenarios offer alternative approaches for practical activity: Scenario 1 (the practical test item) presents a method for incorporating practical mathematical activity into an otherwise conventional test situation. Scenario 2 (sampling) presents a method for continuously monitoring practical mathematical activity.

Scenario 1: The Practical Test Item

Topic: Volume

You will be told when to attempt this item.

Item: On the bench at the side of the classroom you will find a range of mathematical tools: a ruler, some squared paper, a ball of string, scissors, calipers, a measuring cylinder and container of water, a calculator, some centicubes, and some drawing equipment (a compass, dividers, and set squares). You will also find three objects: a wooden cube, a metal cylinder, and an irregular stone.

Using any of the equipment provided or by any other means*, do the following.

a. Find the volume of each of the three objects as accurately as possible.
b. Express your answer in appropriate units.
c. In no more than half a page describe the method you used in each case (diagrams may be used).

*If you think you need additional equipment, please ask for it.

Scenario 2: The Sampling Approach to Practical Assessment

Topic: Skills acquired in any computer-based context (such as statistical methods or geometry using The Geometer's Sketchpad)

Procedure

1. At the beginning of each week, the teacher distributes a list of six skills to be assessed the following week. It may be that new content for that week only accounts for four skills, in which case the teacher includes an additional two skills from previous weeks.
2. Each week, during instructional time when students are working independently, the teacher approaches each student individually and rolls a die. The student must demonstrate competence in whatever skill corresponds to the number on the die.
3. The teacher records the skill demonstrated and the degree of success on a classlist. (A proportion of the student's final grade can derive from this weekly sampling of practical skills.)

Teachers who have used this approach report high student motivation to acquire all six skills each week, as well as student satisfaction with the fairness of the sampling method.

Practical tasks are one example of the need for consistency between type of instructional activity and type of assessment activity that should be applied to all forms of mathematical activity. The two task characteristics that might restrict the authenticity of a student's response on any task are *communication* and *context.* We discussed modes of communication in the section Choosing the Right Task. The important point to recognize here is that some modes of communication are not tenable in a test situation.

With regard to mathematics tasks that include contextual detail, there will be some students to whom the context does not appeal or is simply foreign. Even among those students who find interest in a problem's context, there will be some for whom the demands of language or contextual detail constitute an excessive cognitive load. For this reason, providing diversity across different practical or real-world contexts is not enough. There must also be diversity in the *type* of context and in the *degree of detail* needed to specify the task. We need to ensure that we employ sufficient diversity of contexts and context types to minimize the disadvantage caused to any individual by a particular context. In a test situation the problem is heightened because of the limited range of contexts that can be sampled.

Even in a test format, however, it is possible to include items relating to a particular concept or skill that access at least three distinct context types. Satisfactory responses from a student in all three contexts would provide strong evidence of understanding. One of these contexts might involve a form of practical activity and one a real-world applications task, while a third addresses the mathematical abstraction. For example, consider the following item on bar graphs.

Physical representation
Use only the packet of candies provided to construct a bar graph.

Real-world context
There are thirty students in a mathematics class. The class constructs a bar graph of the number of siblings (brothers and sisters added together) of all class members. What might this graph look like? Draw such a graph and label it appropriately.

Mathematical abstraction

a. What might this be a graph of?
b. Label the graph appropriately.
c. What information is contained in your graph?

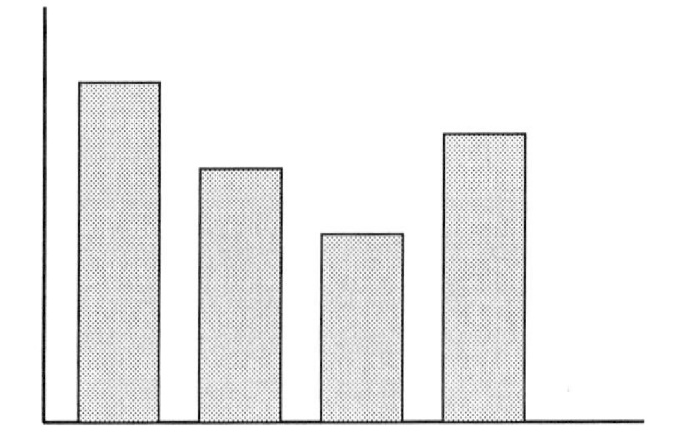

Student-constructed test items

One of the goals of constructive assessment is to make assessment more educational by making more effective instructional use of the time taken up with an assessment event. The class time and the teacher time involved in creating and administering a formal test are significant. How can we fully utilize the educational potential of this expenditure of time? Consider the following scenario.

1. *The review lesson.* Students enter the classroom and are organized into groups of four. Each group is given the following instruction: "Develop five tasks that you think would fairly assess the content covered in the topic we have just completed." To give the exercise credibility, the teacher undertakes to include in the class test at least one item from each group. The teacher retains editorial rights and may change the wording of a question in the interest of clarity, but the general direction of the question is retained.

   To complete this activity students must (a) partition the content of the topic into five sections, each of which can be assessed with a single item and (b) decide on the level of difficulty appropriate for each item. (Teachers who have tried this approach have reported that student-constructed test items are consistently harder than those the teachers would have devised.)

2. *Compiling the test.* The teacher retains the responsibility for selecting the test items from among those generated by the students. It may be necessary to insert one or two teacher-created items to generate a test that adequately represents the topic, but if at all possible, the teacher should strive for including only items created by the students.

3. *Administering the test.* The test is administered like any other test. Teachers report that student interest is high. Students care about which of their items were included in the test. It is also more difficult for students to see the test as an inappropriate or arbitrary imposition by the teacher. These are the items that they and their classmates feel constitute a fair assessment of the content.

4. *Correcting the test.* Some teachers ask student groups to submit solutions and grading schemes with their five questions. Others feel this step reduces student enthusiasm for the activity.

Decisions like these are best left to your own judgment and your feeling for the likely benefits to the class and to you.

5. *Returning the corrected tests.* Often, one of the least effective lessons in a mathematical classroom is the lesson in which corrected tests are returned to students. Most teachers feel obliged to discuss each test item in a final attempt to review the topic or course content. Quite commonly, an important opportunity to assist students in synthesizing their mathematical knowledge and in learning from the test experience is lost because students' interest is not engaged by this type of review. In contrast, teachers who have used student-constructed test items report a high level of student interest in this activity. Students want to know how successful the class was on their particular item and which groups of students were responsible for the more difficult items. Here is an opportunity for the teacher to move the locus of authority visibly in the direction of the students: "Anne, your group was responsible for item 9. What do you think would have been a good response to your question?"

Many teachers have remarked that the process of writing test items is more than just motivationally valuable for students. Many feel that the task of constructing test items is a much more effective way to review a topic than the more conventional summary lecture. And we need not restrict this activity to the construction of test items. Students will also benefit from the experience of developing more elaborate tasks for completion by their peers.

Group assessment

There are at least two compelling reasons to assess student work in groups:

- Many employment situations require that people work in teams, and students will benefit from developing the skills of effective group work.
- The opportunity to learn and to perform mathematically as a member of a group may be both a preferred learning mode and a preferred mode of communication for some students. In other

words, they may be best able to display their mathematical understandings as a member of a group.

In the case of truly collaborative groups, it is very difficult to assess the relative significance of the contributions of individuals. For this reason the most common approach has been to assess the work of the entire group as a single product. Over time a student can perform mathematically as a member of several different groups. An individual profile can then be constructed from the groups' assessments and from comments on individuals in the teacher's annotated classlist.

Teachers who use groups in mathematics classrooms tend to ask students to produce a written group report, to do a group presentation of their findings in front of their peers, or both. The grading of a complex performance like a written group report is best dealt with by using the Eight-Step Development of Mathematics Performance Criteria outlined in Part 3 (page 70).

For group presentations, one method of grading is as follows.

Step 1: The teacher leads a class discussion prior to the group presentations, focusing on the question "What are we looking for?" The result of this discussion should be three (or more) performance criteria. It is important that the teacher and all members of the class share a similar understanding of the established performance criteria.

Step 2: A group makes its presentation to the teacher and the rest of the class, perhaps followed by a brief time for the class and the teacher to ask questions of the group members.

Step 3: The teacher asks the class to identify up to three significant things they feel they learned from the group presentation. It is both important and useful in this discussion to distinguish the detail of a group presentation from the "big ideas" that were presented.

Step 4: The group submits a brief written report on its own presentation. The focus of this report is the group's answers to the questions "What were the strengths of your presentation?"

> and "What would you do differently next time?" This group report offers students the opportunity to demonstrate that they have learned from the experience. As such it should be accorded some weight in any grading process. Alternatively, the teacher may require such a written report from each student as an additional performance requirement.

Some teachers may feel that group work is a powerful instructional method and a useful learning mode, but one that does not require assessment. They may want to treat group work simply as an important component in the students' educational experience. Some teachers may wish to provide feedback to students on the quality of their group presentation but not record a grade for the presentation. For those teachers who feel a grade is appropriate (or required), the box Grading Group Presentations provides a suggested structure for grading group presentations.

Grading Group Presentations

Teacher. Each of three performance criteria is rated on a three-point scale.

Criterion 1 (e.g., clarity)	High / Medium / Low
Criterion 2 (e.g., level of math)	High / Medium / Low
Criterion 3 (e.g., preparation)	High / Medium / Low

Peers. Class discussion aims to identify up to three significant things the class feels they learned from the presentation.

Group. Group members submit a combined report on the strengths of their presentation and on those things they would do differently the next time.

Final grade. The teacher and peer assessments are combined to generate a "preliminary grade" by some procedure determined by the teacher in consultation with the class. This preliminary grade may be adjusted by up to one grade point in light of the group's evaluation of its own presentation. A group whose presentation had significant flaws is thus given an opportunity to suggest how these flaws might be overcome and to receive recognition for these ideas in the grading scheme.

Student self-assessment

Control and ownership of assessment events has traditionally remained in the hands of the teacher or some external authority. Central to the concept of constructive assessment, however, is the sharing of responsibility for assessment between teacher and student. Given the significance attached to assessment, student involvement in the assessment process provides an excellent opportunity to demystify assessment, to further integrate assessment into instruction, and to shift some of the responsibility and workload associated with assessment from the teacher to the students. We have already seen an example of this shared responsibility in the discussion of student-constructed test items. Student self-assessment is another step in this direction, one that has the additional benefit of tapping students' feelings and attitudes as well as their cognitive processes.

In one type of student self-assessment, students might be asked to respond every two weeks to items such as "What is the biggest worry affecting your work in math at the moment?" "Write down one particular problem that you found difficult," and "What was the best thing that happened in math in the last two weeks?" These items require students to reflect on their learning and to articulate the consequences of that reflection. A sample response sheet for one such procedure (IMPACT) is provided in the accompanying box.

Actual responses to questions like these have included the following:

"Write down one particular problem that you found difficult."

> Algebra a bit, because I don't understand why we don't just use numbers. It would be simpler.

"Write down one new problem that you can now do."

$$\frac{1}{3} \bullet \frac{4}{1} = \frac{4}{3} = \frac{1}{12}$$

"How might we improve math classes?"

> Have less work and more learning.

Such responses offer insights into student perceptions, conceptions, and understandings (or misunderstandings) that have not been accessed through more conventional modes of assessment.

IMPACT
(Interactive Monitoring Program for Accessing Children's Thinking)

Name:
Class:
Teacher:
Date:

- Write down the two most important things you have learned in math during the past month.
- Write down one particular problem that you have found difficult.
- What would you most like more help with?
- How do you feel in math classes at the moment? (Circle the words that apply to you.)

a. Interested	b. Relaxed	c. Worried
d. Successful	e. Confused	f. Clever
g. Happy	h. Bored	i. Rushed

j. Write down one word of your own. ____________________

- What is the biggest worry affecting your work in math at the moment?
- How might we improve math classes?

Teachers who have used IMPACT report surprise at the frequency with which students refer to personal feelings. Of course, many educators have always acknowledged the significance of student attitudes and feelings as a factor in effective learning, and teachers appear to value student confidence and interest in the subject:

> Interviewer: What do you want your students to get out of your class?
>
> Teacher: That they feel good about themselves, their math abilities; that they're able to problem solve, they see the big picture; that they're confident about geometry, do well on the SAT, are prepared for other math courses.

This teacher shares with many others a feeling that it is just as important to nurture students' attitudes as it is to promote the learning of content. This concern has repeatedly been supported in the research literature. As Gammage (1985) remarks, "It may be in the *emotional* context of the curriculum that the teacher can most modify, alter, or stimulate reactions to learning." And if there were any doubt about the importance of student affect it would be dispelled by students themselves:

> I couldn't understand her. And she used to get in a big huff 'cause I kept putting my hand up asking her how to do things, and then after that, um [long pause] . . . And she used to say I weren't listening. And after that I just used to muck around all the time. 'Cause I didn't like it. I failed.

If we value "the emotional context of the curriculum," then we need to make room for it in our assessment practices.

Assessment is an act of communication, a continuing conversation whose subject is as much the student's self-esteem as it is the student's knowledge. As I have been emphasizing, constructive assessment starts with a concern for the student. Our methods of assessment are one way of saying, "I care about your learning." Conventional assessment has always focused on cognitive outcomes, but we also need to collect information about the affective outcomes of our teaching, such as students' motivation and engagement in mathematics. The most organized and competent teacher can fail by ignoring students' attitudes, and the most carefully constructed curriculum can fail if individual students do not want to learn or do not feel that their efforts are valued by the teacher.

One way to assess what we value is to attend to student affect in an explicit and structured way in our assessment. This does not mean grading student attitudes, but it does mean collecting and exchanging information about such things as students' perseverance, enthusiasm, self-esteem, interest, enjoyment, motivation, anxiety, confidence, and pride in their work.

Here again I can only emphasize the importance of aligning assessment work with instructional goals and values. Some curriculum initiatives appear motivated in part by a desire to promote positive attitudes in students and to maximize their engagement and interest in mathematical activity. Such programs assume a link between academic achievement and attitude, and the existence of this link appears to be substantiated by research. Why, then, do we find little recognition being given to student affect in teachers' assessment strategies? Perhaps it is because assessment has traditionally been so closely linked to grading, and few teachers or curriculum developers would consider grading a student's enthusiasm. But assessment, I have argued, is not the same as grading. And as good teachers we should attend to more than just academic content.

Student journals

Another technique for accessing students' thinking and feeling is the use of journals. Journals have the particular virtue of developing in students a routine of regular reflection on their mathematical activity and learning. In one school, students at all year levels completed a journal entry after each mathematics lesson (Clarke, Waywood, and Stephens 1994). As a first introduction to journal writing, seventh grade students were supplied with a book in which each page was divided into three sections:

- What we did
- What I learned
- Examples and questions

Student journal entries were assessed in terms of quantity, presentation, and sophistication of expression. Dialogue, the most sophisticated mode of expression, was associated with the emergence in the student's writing of a personal voice, denoted by the first person singular; of a speculative ("What if . . . ?") or interrogative ("Then, how could the square . . . ?") style; and of the presence of constructs that were clearly the student's original creations. In the following journal excerpt, a student discusses the idea of "quality of operation."

> Another thing, transposition and substitution really shows you the quality of operations. Like division, it is sort of a secondary operation, with multiplication being the real basis behind it. This ties in with my learning about reading division properly [in previous pages], that is, fractions are different forms of multiplication. So I guess that's like rational numbers (Q) are like a front for multiplication, an extension of multiplication. Which came first, multiplication or division? It would have to be multiplication. They are so similar, no, that's not

what I mean. I mean, they are so strongly connected. But it's like division does not really exist, multiplication is more real. The same with subtraction. Addition and multiplication are the only real operations.

This student is expressing a highly personal and creative view of mathematical operations. "Quality of operations" is not something discussed by the teacher; it is the student's own way of thinking about a mathematical idea. Here the use of journals is important both for what it reveals about the student and because it represents an all-too-rare opportunity for mathematical speculation and creativity.

The effective use of student journals does require a commitment on the part of the teacher to read the journals regularly and to comment on the entries in sufficient detail to facilitate students' progression from description or recounting through summary to dialogue. These comments should be made at least as frequently as once every three or four weeks. The box Criteria for Assessing Student Journals shows one set of criteria that might guide a teacher's comments on a student's journal writing (Clarke, Waywood, and Stephens 1994).

Criteria for Assessing Student Journals

A. Quantity of work

1. Frequency: Is it done after every lesson?
2. Volume: The amount of work done can be taken as a measure of both ability and enthusiasm.

B. How well is it used?

1. Is the work summarized, and do the summaries indicate the development of note-taking skills?
2. Is the journal used to collect important examples of procedures and applications?
3. Are errors or tasks identified and discussed?
4. Are there signs of involvement with the work, original or probing questions, a willingness to explore, and so on?
5. Is the student learning to "dialogue," asking his or her own questions and then setting about to methodically seek an answer and present the investigations logically?

If we are to initiate conversations with our students about their enjoyment of mathematics, their anxieties, and how they see themselves as mathematics students, we need assessment strategies and information on which to build the conversation. Happily, these strategies already exist. Both IMPACT and student journals have the capacity to reveal students' affective response to mathematics and the mathematics classroom.

Student portfolios

A final technique for monitoring good practice, the use of student portfolios, is rapidly gaining adherents. Student portfolios offer students the chance to demonstrate the evolution of their mathematical knowledge and performance over the duration of a topic or a course. The power of a portfolio lies in its demonstration of growth or development in a student's performance and in the clarity of communication it offers for discussions of progress between teacher and parent, teacher and student, or parent and student.

There are many ways to implement the use of student portfolios. The sample scenario shown in the Scenario for Portfolio Use box is one recommended approach.

Variations on this scenario could include other performance types, including student self-assessments. One additional detail, employed by many teachers, is to require that students attach a brief statement to each portfolio item in which they complete the sentence "I have included this piece of work in my portfolio because it shows . . . ."

In contrast to the purpose of a developmental portfolio, which is to show growth, another purpose of portfolios is to show accomplishment. The model I have described can be used to serve both purposes. Another useful model, one that focuses on accomplishment, is provided by the New Standards assessment system (New Standards 1996). New Standards is a partnership of urban districts and states (including, among others, California, Kentucky, and Vermont) working collaboratively to build an assessment system whereby they can measure their students' progress in meeting national content standards at levels that are internationally benchmarked.

The New Standards assessment system has three interrelated components: performance standards, an on-demand examination, and a portfolio system. Its unique feature, distinguishing it from other consortia of states and from

Scenario for Portfolio Use

1. *Establishing representation.* * The teacher establishes what the key components of a student portfolio are to be, perhaps through discussion with the class. One approach is to require the inclusion of five different types of mathematical performance.

a. An investigative project (a substantial piece of work requiring at least one week for completion)
b. A substantial nonroutine problem-solving task (requiring at least two hours for completion)
c. A set of three contextually diverse performances related to the same concept or skill (one physical representation task, one real-world task, and one abstract task)
d. A worksheet of five content-specific, open-ended tasks (Sullivan and Clarke 1991)
e. A test

The teacher then structures students' mathematical activity throughout the year in such a way as to generate in a regular fashion each of these five performance types.

2. *The first five pieces.* The first items to be included in the portfolio are the student's first attempts at each of the five performance types. These first five pieces are dated and remain in the student's portfolio throughout the year.

3. *Additional portfolio items.* One additional example of each of the five performance types may be added at any time during the year. The purpose of adding this second example should be the demonstration of improvement in this type of performance. This new item becomes the "benchmark" item, indicative of the best possible performance of this type to date. At no time can there be more than two examples of any one type in the portfolio. Because the first sample item of a particular type remains in the portfolio all year, once a second performance of that type has been included, any additional items of that type can only be included by replacing the existing benchmark item. Following this procedure, at no time will a portfolio contain more than ten pieces of student work.

* Performance types a and b require tool selection, type c requires application, type d offers the opportunity to demonstrate understanding, and type e is concerned with tool possession.

commercial publishers who are developing performance assessments, is that it is based on explicit performance standards.

The performance standards are derived from the national content standards developed by professional organizations (the NCTM *Assessment Standards,* in the case of mathematics) and consist of two parts:

- *Performance descriptions.* Descriptions of what students should know and of the ways in which they should demonstrate the knowledge and skills they have acquired in the four areas assessed by New Standards—English language arts, mathematics, science, and applied learning—at elementary, middle, and high school levels
- *Work samples and commentaries.* Samples of student work that illustrate the meaning of the performance descriptions, together with commentary that shows how the performance descriptions are reflected in the work sample

The on-demand examination, called the *reference examination* because it provides a point of reference to national standards instead of national norms, assesses those aspects of the performance standards that can be assessed in a limited time frame under standardized conditions. The reference examination stops short of being able to accommodate longer pieces of work—reading several books, writing with revision, performing investigations in mathematics and science, and completing projects in applied learning—that are required by New Standards performance standards and the national consensus content standards from which they are derived.

The portfolio system complements the reference examination by providing evidence of achievement of those performance standards that depend on extended pieces of work. Modeled after Advanced Placement portfolios, the system calls for portfolios organized in exhibits that have focused purposes and clear criteria for judgment. Each exhibit is composed of one or more entries; entry slips are provided that tell students exactly what is required and how the entry will be evaluated. The following chart summarizes the components of the portfolio.

New Standards Portfolio System

Exhibit	Entries	Comments
Conceptual Understanding	Four entries, one for each major area of mathematics • Numbers and operations • Geometry and measurement • Functions and algebra • Statistics and probability	To demonstrate conceptual understanding, students provide evidence that they can use a concept to solve problems, represent the concept in multiple ways (through numbers, graphs, symbols, diagrams, or words), and explain it to someone else.
Problem Solving	Four pieces of work that, taken together, show the full range of problem solving • Formulation • Implementation • Conclusion	Problem solving is defined as using mathematical concepts and skills to solve nonroutine problems that do not lay out specific and detailed steps to follow.
Putting Mathematics to Work	At least one large-scale investigation or project each year (and, over the course of high school, investigations or projects drawn from at least three of the types shown at right)	Types of investigations: • Data study • Model of a physical system or phenomenon • Design of a physical structure • Management and planning analysis • Pure mathematics investigation
Skills and Communication	Two entry slips that ask students to locate work in the other exhibits. If earlier entries are insufficient to demonstrate skills and communications, additional pieces of work may be included here.	Lists of skills and communications are presented on the entry slips.

Many of the benefits of developmental portfolios are realized in standards-based portfolios. Whether they are used to show development or accomplishment, portfolios focus students on their responsibility for producing work, and they focus conversations between teachers and students on the students' work. They also put the standards or expectations directly into the hands of students; unless we communicate explicitly with students about what we expect, it will be very difficult for them to keep up their end of the Didactic Contract.

The various strategies we have discussed in Part 2 are in no way a comprehensive selection of assessment activities. They do, however, represent the strategies that teachers have endorsed most consistently as useful and manageable. Each has been adapted by many teachers to better suit the needs of their school and classroom contexts. If you are considering expanding your repertoire of assessment practices, the advice of the teaching community is clear: Start small and go slow. Select one new strategy and commit yourself to its use for at least one school term. If it becomes part of your teaching routine, you might consider adding a further strategy. If it does not meet your needs, try a different approach.

Progressively, you will find yourself in possession of a rich body of information on your students and on your teaching. The question that remains is, How can we best use all this information? Part 3 addresses this question in detail.

PART 3

INFORMING GOOD PRACTICE BY TEACHERS, STUDENTS, AND OTHERS

Assessment is about the exchange of information. If your assessment program is to play a constructive role in your students' learning, then the communication of assessment information must be clear and purposeful, and it must inform action. With regard to clarity of communication, it may be that your students have something useful to say.

How much of the responsibility for assessment can you usefully share with your students? A teacher who has collected many diverse samples of a student's work and who has observed the student and engaged in conversation with the student about mathematics may see no need for a formal test. However, if the Didactic Contract has established that it is the student's responsibility to demonstrate understanding, then the teacher might propose that the student be awarded a particular grade with the proviso "If you don't like the grade I have proposed, you can show me I am wrong by doing this test." If the test has been compiled from student-generated items, then the entire process becomes optimally constructive.

The principal focus of constructive assessment is always, How can I make this assessment a better portrayal of the student's learning, and how can the information generated in this assessment usefully inform someone's action? The simple rule for effective communication is, Ask the audience. With regard to communicating assessment information, ask yourself, ask your colleagues, ask parents, and ask your students the following:

- What is it that your students most need to hear regarding their learning?
- What is it that their parents most need to hear?
- What is it that other teachers might usefully be told about each student?
- What documentation does the school require?
- What actions do you hope to inform through your assessment?
- What form of presentation of assessment information will most likely prove useful to each of the recipients?

As a test case, ask yourself the questions, What is the difference in information content between a grade and a portfolio of student work? Whose actions are likely to be usefully informed by each?

Contemporary assessment recognizes the inadequacy of the "Assessment as measurement" metaphor. Our goal is now "Assessment as portrayal."

Grading Alternatives: To B or Not to B

> I must be reading the wrong books on alternative assessment. None explain how to interpret the assessments into letter grades. (Mathematics teacher, San Francisco Bay Area)

> How can I explain to my students what is expected in this new mathematics performance assessment? I'm not even sure I know myself. (Teacher at an assessment workshop)

Grading makes sense only if both the teacher and the student agree on what constitutes good mathematics performance. The student must understand the meaning to be associated with each grade. To this end, both the purpose and the process of the assessment scheme should be shared with students and parents. Further, students themselves can and should be involved in establishing the criteria by which their mathematical performance will be graded. In this section we consider both the purpose of grading and the process by which it can be most usefully done.

What purpose is served by grading a student's performance in mathematics? Grading, as I noted earlier, is not assessment; it is simply one means of coding assessment information. Like most coding schemes, grading serves to condense and to categorize. Yet it is apparent that a single score or grade cannot usefully summarize the assessment information collected from classroom observations, questioning, group reports, tests, or responses to an open-ended task. Such a grading process sacrifices precisely that detail that might contribute most constructively to the subsequent actions of teacher, student, or parent.

Assessment criteria are an attempt to structure the grading process—to make it a meaningful categorizing system rather than just a one-dimensional measurement. Four points are important to note in this connection.

- All grading is an attempt to simplify complex information.
- All grading is selective.
- Grading quantifies (or categorizes) a specific predetermined facet of mathematical performance.
- All grading discards information.

Changing the function of assessment from "measurement" to "portrayal" involves the rejection of any single criterion or grade as an adequate summation of students' mathematical performance. Instead, we identify a set of criteria that describe and categorize those aspects of student mathematical performance we think are important. Curriculum documents like the NCTM *Assessment Standards* help to identify the big ideas and important skills. Such a set of criteria may provide, in combination, a succinct coded portrayal of a student's performance, one that is easily recorded and easily communicated.

Developing and sharing these criteria depends upon achieving, with our colleagues and our students, a shared understanding of the goals of our classrooms. Some time ago, a teacher at an assessment workshop in the San Francisco Bay Area posed a fundamental issue:

All grading discards information.

> David, I think my problem is that my kids aren't clear on what good mathematics looks like, and to be honest, I don't think I am all that clear on it either.

This teacher's candid observation highlights the fact that many of the new tasks finding their place in mathematics classrooms challenge the existing Didactic Contract. An example of such a task is "Which is the better fit: A square peg in a round hole or a round peg in a square hole?" (Schoenfeld 1985). Many students feel that such a task makes unreasonable demands; many are confused about what is required; and many teachers are unsure how to assess the diverse responses such a task produces. Such tasks stimulate rich and complex mathematical activity and offer significant insight into students' ability to use the mathematical tool skills they possess. However, introducing such tasks into the classroom requires renegotiating the Didactic Contract. Essentially, we must establish a consensus within our classrooms about what constitutes quality mathematics.

How, then, do we bring our students to share our vision of what quality mathematics looks like? Classroom field-testing and conversations with many teachers have led to the following strategy for establishing the terms of the Didactic Contract in order to prepare the way for using the new diversity of mathematical tasks in instruction and in assessment.

Establishing the criteria for quality mathematics performance

Constructive assessment seeks to optimize student involvement in the assessment process. Through such involvement students can develop the following:

- An understanding of the characteristics of sophisticated mathematical performance (at their existing level of competence)
- A familiarity and feeling of ownership in relation to the purposes and practices of assessment
- A critical facility that enables them to judge the quality of their own mathematical performances and those of others

Consistent with this goal is the strategy of involving students in the development of rubrics for assessing mathematical performance. A key feature of this approach is that, while the teacher will have a set of criteria in mind, the starting point is the set of criteria that the students identify. The teacher then clarifies and adapts these criteria in consultation with the class until an appropriate set that everyone understands is established. In the scenario that follows, criteria are developed through discussion in small groups of samples of other students' problem-solving attempts.

1. Groups of teachers collect a variety of tasks reflecting the contemporary emphasis on task diversity, together with sample student performances in response to each task. Ideally, these sample performances should include one good performance, one poor performance, and one performance of average quality. Banks of tasks and corresponding sample student performances are compiled for use by the group.

2. From the bank of available tasks, the teacher selects one of a type that is not too dissimilar from those within the students' recent classroom experience. This task is distributed in class, and students are given an appropriate amount of time to attempt it. The teacher emphasizes that this is a new sort of mathematical task and that the purpose of the exercise is to familiarize everyone with the new task type. An example of a task is "Which is the better fit: A square peg in a round hole, or a round peg in a square hole? Explain your answer." All students are encouraged to "do their best" because some students will find the new tasks quite threatening. The use of small groups is recommended. In particular, the teaching strategy "Think, Pair, Share" may be useful. In this strategy, students work alone to clarify their understanding of the task and to identify relevant facts and skills, then in pairs to solve the problem, and then in groups of four to share their solutions and attempt to reach consensus before a discussion of the task by the class as a whole.

3. After a brief discussion of the task, the teacher distributes to each group samples of other students' attempts at the task—one good, one poor, one average—but none labeled or graded in any way

that might indicate their relative merit. Students are asked to select the response they feel is the best, then to select the worst, and finally to justify their choices by identifying what particular qualities of the responses distinguished the good response from the poor response. The teacher records these response attributes, as these will become the reference point for future discussion of class responses to this task type.

4. Teachers with an obligation to grade student performance may take the additional step of leading the class in determining a rubric, based on the previous discussion, that then becomes the basis for grading future attempts at tasks of this type. Students can be encouraged to apply the new rubric to their own responses, with the acknowledgment that this first exercise was a familiarization lesson and that the student grades will not be included in any significant assessment.

The procedure just outlined can be implemented each time a new task type is introduced. In each case, consensus should be reached between students and teacher concerning the terms of the new Didactic Contract.

One of the keys to effective communication is involvement. If students and parents can be involved in the assessment process, then it is more likely that they will understand both the process and the information it generates. The box Eight-Step Development of Mathematics Performance Criteria presents this process as a series of simple steps (see next page).

The assessment or grading scheme that results from this process not only has the endorsement of all students as well as the teacher, but also is far more likely to be used and interpreted with understanding. It is also likely that the process of negotiating the wording of the assessment scheme will be educationally valuable for the students, requiring them to think carefully about the characteristics of quality mathematical performance.

Reporting techniques

How should we communicate the results of our assessment efforts? Given the limitations of grades, one option is to put more emphasis on student portfolios.

Eight-Step Development of Mathematics Performance Criteria

Step 1: Pose an open-ended task suitable for the unit and let the students do the task.

Step 2: Distribute three samples of other students' attempts at the same task and ask the class (in groups of four or in pairs) to identify the best and worst attempts and to find reasons for their decision.

Step 3: Share all the things that students feel distinguish a good solution from a bad one, add some of your own as necessary (with the endorsement of the class), and generate a list of criteria for good performance.

Step 4: Develop as a class a grading scheme by which the criteria can be translated into a grade or score for the task.

Step 5: Ask the students to apply the new criteria to their solution to the problem and to make evaluative comments on their own work.

Step 6: Ask the students to apply the grading scheme to their solution and generate a grade for their work.

Step 7: Have the students submit their solution with their evaluative comments and grade.

Step 8: "Second-grade" the work to check for shared understanding of the agreed criteria and consistency in their use.

A student's portfolio can be a very powerful focus for constructive conversations at parent-teacher meetings. Used in this way, portfolios can significantly advance the primary goal of assessment—to inform subsequent action by students, parents, and teachers.

School report cards, of course, are intended to be a succinct summation of students' performance and progress. There have been many attempts to make report cards more informative. The box Sample School Report Card Format displays an example that represents a compromise between elaborate descriptive

Sample School Report Card Format

Name: ______________________ **Tutor group:** ____________

Subject: Mathematics 8 **Semester 1**

The Year Eight program aims to continue the development of student numeracy and awareness of space and shapes, and begins to consolidate algebraic skills and use them to model real situations. Knowledge of concepts associated with chance events and the analysis of data continues to be developed. Students build on their earlier project experiences by preparing an entry in the Mathematics Talent Quest. Problem solving is encouraged through work on Task Center activities, and work involving computers and calculators is integrated throughout the semester's units.

Work requirements		Graded assessment	
1. Skills and standard applications	☐	1. Graphs of experiments	☐
2. Problem solving and modeling	☐	2. Integers, Cartesian coordinates	☐
3. Project work	☐	3. Correlation	☐
S = Satisfactory completion N = Unsatisfactory completion		4. Pyramids, compass constructions	☐
Attitude and management		5. Ratio and rates	☐
1. Acceptable class behavior	☐	6. Bookwork	☐
2. Ability to work independently	☐	7. Project: Mathematics Talent Quest	☐
3. Completion of set tasks	☐	8. Problem solving: Math Task Center	☐
4. Ability to work cooperatively	☐		
5. Motivation	☐		
6. Participation in discussion	☐		
H = High M = Medium L = Low			

Descriptive comments: ______________________

Teacher's signature: ______________________

detail and grades in an attempt to provide a more complete portrayal of the student's mathematics performance.

Enacting the Constructive Assessment Agenda

Through the tasks and methods we employ and through the continual exchange of assessment information within the school community, our assessment should *model* quality mathematics, *monitor* associated mathematical learning, and *inform* the school community concerning the level of success in teaching and learning mathematics in ways likely to facilitate further learning. These three functions—to *model,* to *monitor,* and to *inform*—are fundamentally linked. A comprehensive assessment program must address all three. Of course, changes to the curriculum should involve changes to each assessment function. Above all, we should work to ensure that all components of our assessment program will genuinely assist everyone involved in improving instruction, learning, and the support provided by parents and the school.

No one has ever gotten any taller just by being measured.

Assessment contributes constructively to our educational efforts when it portrays the student's learning with optimal richness and accuracy. As I have emphasized, some forms of assessment may misrepresent particular students. A student's test performance may be unrepresentative for reasons such as excessive linguistic demands, unfamiliar problem contexts, short-term personal problems, test anxiety, or inadequate time for completing the task. Similarly, oral presentations of group problem-solving reports may misrepresent individual student understandings for reasons such as lack of verbal fluency, the stress associated with public speaking, uneven sharing of responsibilities in the problem-solving process, or the difficulty of adequately displaying mathematical methods or meanings through this medium of communication. It follows that students should be offered a variety of modes of communication through which to display their mathematical understandings. Furthermore, allowing students to choose a preferred mode of assessment may increase the likelihood that the assessment is both rich and accurate.

A school or a mathematics department seeking to enact the constructive assessment agenda, as outlined in this book, can do so in many ways. Contemporary mathematics teachers have an impressive repertoire of assessment strategies from which to choose:

- Tests (including practical tests and tests composed of student-constructed items)
- Annotated classlists
- Student self-assessment techniques
- Portfolios (comprising a diversity of mathematical performance types)
- Student mathematics journals
- Investigative projects
- Problem-solving tasks and a wide range of open-ended tasks

If implemented in line with the principles and examples outlined in this book, all of these strategies can make a constructive contribution to teaching and learning. The difficulty for the teacher is how to integrate these new approaches into a daily routine that is already extremely demanding. One teacher expressed a common concern:

> If alternative assessment could be done without a ton of extra paperwork, I think I would be happy to try something besides tests and quizzes.

A major purpose of this book has been to begin addressing this concern by providing a practical approach to expanding our mathematics assessment. Every recommended strategy has been used by classroom teachers with success. But we must be selective in choosing our strategies. Not every strategy is relevant to every topic, every grade level, or every student. Furthermore, no matter how ambitious we might be with regard to the assessment of our students, we work in real schools where many constraints restrict our actions. So let me restate that the goal of our efforts is not *more* assessment but *better* assessment. We need to be highly selective in our assessment strategies and clear about the benefits of any strategy we choose to employ. If we are considering administering a test, we must ask the questions, For what purpose? Whose actions will be informed by the results of this test? If we are considering introducing portfolios or student self-assessment, we must ask, What will be the additional demands on class time and my personal time? Does the information I will gain justify the cost in time and effort?

By the same token, a school or a department considering expanding its assessment practices would profit by adopting the motto Start Small, Go Slow. One scenario for change within a mathematics department might involve pairs of teachers at different grade levels selecting one new assessment strategy to field-test over one term. At this experimental stage a teacher should not try more than one of these ongoing strategies with a particular class. Introducing new assessment techniques gradually will help ease teachers' workloads, focus student and teacher effort on the new strategy more effectively, and honor the need to renegotiate the Didactic Contract as a consequence of introducing a new form of assessment.

Once teachers have become familiar with the demands and the rewards of the various assessment strategies, then teams of teachers at a given grade level can begin to integrate new strategies into their mathematics program (for sample programs, see Appendix B). These new strategies need not all be added to existing methods. Indeed, some existing practices might be replaced or modified.

Of course, significant change is never easy, but neither is anything worthwhile. The potential for assessment to contribute constructively to every facet of the curriculum has seldom been realized in practice. I firmly believe that once we commit to a program of truly constructive assessment, we will find that the rewards far outweigh the costs.

Not more assessment
but better assessment.

REFERENCES

Brousseau, G. (1986) "Fondements et methodes de la didactique des mathematiques." *Recherches en didactique des mathematiques* 7(2) 33–115.

Clarke, D.J., and Helme, S. (1994) "Context as Construct." *Contexts in Mathematics Education—Panel Discussion Papers,* B. Atweh, ed. Kelvin Grove, Queensland: Mathematics Education Research Group of Australasia (MERGA) 1–10.

Clarke, D.J., Waywood, A., and Stephens, M. (1994) "Probing the Structure of Mathematical Writing." *Educational Studies in Mathematics* 25(3) 235–250.

Gammage, P. (1985) "Perspectives on the Personal: Social Psychology and Education" (inaugural lecture at the University of Nottingham School of Education, 7 November 1985).

National Council of Teachers of Mathematics (NCTM). (1989 and 1995) *Curriculum and Evaluation Standards for School Mathematics.* Reston, Virginia: NCTM.

New Standards. (1996) *New Standards Performance Standards.* Washington, DC: National Center on Education and the Economy.

Schoenfeld, A. (1985) *Mathematical Problem Solving.* London: Academic Press.

Sullivan, P., and Clarke, D.J. (1991) "Catering to All Abilities Through 'Good' Questions." *Arithmetic Teacher* 39(2) (Oct) 14–18.

APPENDIX A

THINGS TO SAY TO PARENTS AND CONCERNED OTHERS—AN ANNOTATED BIBLIOGRAPHY

Teachers and administrators will find it very useful to begin assembling a library of reference material to which they can turn for support in discussions with parents, colleagues, and other stakeholders regarding the new assessment agenda. While certainly not comprehensive, the list provided here represents an excellent start on an assessment reference library.

Charles, R., and Silver, E., eds. (1988) *Teaching and Evaluating Mathematical Problem Solving.* Reston, Virginia: NCTM.

A standard reference on the assessment of mathematical problem solving. Perhaps a little dated in its perspective but useful nonetheless.

Clarke, D.J. (1988) *Assessment Alternatives in Mathematics.* Canberra, Australia: Curriculum Corporation.

Stenmark, J. (1989) *Assessment Alternatives in Mathematics.* Berkeley, California: EQUALS Project, Lawrence Hall of Science.

Two books with the same title and the same basic purpose. Both useful "how to" books, both available through NCTM.

Leder, G., ed. (1992) *Assessment and Learning of Mathematics.* Hawthorn, Australia: Australian Council for Educational Research (ACER).

One of the best collections of theoretical and practical writings on assessment in mathematics. Available by writing to ACER, 19 Prospect Hill Road, Camberwell, Victoria 3124, Australia.

Mathematical Sciences Education Board, National Research Council. (1993) *Measuring What Counts: A Conceptual Guide for Mathematics Assessment.* Washington, DC: National Academy Press.

National Academy of Education. (1993) *Setting Performance Standards for Student Achievement.* Stanford, California: National Academy of Education.

Two contemporary standard guides to national policy regarding assessment practice.

National Council of Teachers of Mathematics (NCTM). (1989 and 1995) *Curriculum and Evaluation Standards for School Mathematics* and *Assessment Standards.* Reston, Virginia: NCTM.

Two key documents shaping the direction of the mathematics curriculum into the next century. The recommendations are far-sighted and carry the endorsement of the national professional association. In combination with the Professional Standards document (NCTM 1993), these provide a central reference point for discussions of school mathematics in the United States.

New Standards. (1996) *New Standards Performance Standards.* Washington, DC: National Center on Education and the Economy.

Describes performance standards that build on the NCTM standards and addresses the question, How good is good enough? by including examples of student work that illustrate standards-setting performances.

Schoenfeld, A.H. (1994) "Learning to Think Mathematically: Problem Solving, Metacognition, and Sense-Making in Mathematics." *Handbook for Research on Mathematics Teaching and Learning,* D. Grouws, ed. New York: Macmillan. 334–370.

A state-of-the-art discussion of higher-order mathematical thinking. The Grouws *Handbook* should be on every school library shelf. See also the chapter on assessment by Norm Webb.

Sullivan, P., and Clarke, D.J. (1991) *Communication in the Classroom: The Importance of Good Questioning.* Geelong, Australia: Deakin University Press.

A teacher-friendly monograph on good questioning in mathematics. Available by writing to Deakin University Press, Geelong, Victoria 3217, Australia.

Webb, N., ed. (1993) *The 1993 NCTM Yearbook on Assessment.* Reston, Virginia: NCTM.

Romberg, T., ed. (1992) *Mathematics Assessment and Evaluation: Imperatives for Mathematics Educators.* New York: The State of University of New York Press.

Two useful compilations of contemporary writings on school mathematics assessment.

APPENDIX B

A CONTINUUM OF MATHEMATICAL TASKS

Contemporary mathematics teachers do not lack suitable tasks for assessment and instruction. Rather, the challenge is to make a discerning and informed selection, identifying the task type that is most likely to prompt the desired mathematical performance. The examples that follow represent a continuum from closed, content-specific, multiple-choice tasks to open-ended investigative projects.

Large amounts of money and personnel are being devoted to the development of tasks that meet the requirements of the newly conceived assessment agenda. It is important to identify the range of task types already available. In the listing below, examples of task prototypes are provided in a notional order of increasing complexity. (These tasks are adapted from various United States, Australian, and Canadian sources.)

1. Multiple-choice questions

   Three students each have a probability of 0.8 of getting their assignments completed by the due date, independently of each other. What is the probability that none of the three assignments is completed by the due date?

   a. 0.008
   b. 0.2
   c. 0.512
   d. 0.6
   e. 0.8

2. Enhanced multiple-choice questions

$$\begin{array}{rrr} \square & \square & \square \\ \times & \square & \square \\ \hline \end{array}$$

The five digits 1, 2, 3, 4, and 5 are placed in the boxes above to form a multiplication problem. If the digits are placed to give a maximum product, the product will fall between which values?

a. 10,000 and 22,000
b. 22,001 and 22,300
c. 22,301 and 22,400
d. 22,401 and 22,500

3. Numerical-response question

The first three terms of an arithmetic sequence are $19 - x$, $3x$, and $4x - 1$. Correct to the nearest tenth, the numerical value of the second term of this sequence is ________ .

4. Short answer (explicit cuing)

Four cheerleaders are working on a special routine for a high school basketball game.

a. In how many ways can the four cheerleaders arrange themselves in a row?
b. In how many ways can the four cheerleaders arrange themselves in a circle?

5. Good Questions (open-ended, specific content domain)

The average of five numbers is 17.2.
What might the numbers be?

Good Questions are characterized by their content-specific focus and the opportunity for answers at different levels of sophistication.

Besides mathematical correctness, performance can include providing a single answer, recognizing the existence of multiple answers, listing all answers, or providing a general statement encompassing all possible answers.

6. Extended-answer question (explicit cuing and guidance)

   Steve was fined $95 for speeding in a 55-mph zone. How fast do you think he was going? To answer that, you'd need to know the formula for determining fines. Different communities determine fines in different ways. Where Steve was speeding, they use the equation $y = 3x + 50$, where y represents the amount of the fine in dollars and x represents the number of miles per hour in excess of the speed limit. So, for example, a driver going 10 mph over the speed limit would be fined $3(10) + 50 = \$80$.

   a. Graph the equation $y = 3x + 50$. What does the slope of this line represent? What does the y-intercept mean?
   b. What would the fine be for going 20 mph over the speed limit?
   c. If Steve was fined $95 for speeding in a 55-mph zone, how fast was he going?
   d. Is it possible to be fined $50 for speeding? How could you indicate this on your graph?
   e. Do you suppose this formula would apply to someone going 50 mph over the speed limit? Explain.

7. Extended answer (reduced cuing and guidance)

   Monica and Denee are sitting on the front deck of their cabin. They pride themselves on their estimation skills and are taking turns estimating distances between objects in their surroundings. There are two large redwood trees visible from where they are sitting, one on the left side and one on the right side of their property. Denee claims that the two trees are 180 feet apart, and Monica says they are 275 feet apart.

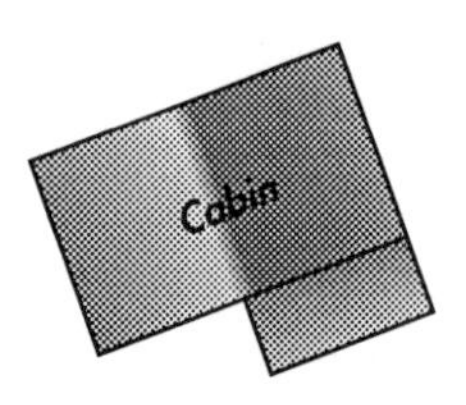

The problem is, they can't measure the distance to see whose estimate is closest because the cabin is located between the trees. All of a sudden, Denee recalls her geometry: "Oh yeah, the Triangle Midsegment Conjecture!"
She runs to get a tape measure, a hammer, and a stake. What is she going to do?

8. Open-ended, extended-answer question (some cuing of method)

   A school mathematics club is designing games for students to play at a school carnival booth. Here is one of the games.

   > Take two ordinary dice of different colors (for example, one white die and one red die). Roll both dice together. The student player wins when the number on the white die is greater than the number on the red die. The math club wins otherwise.

   Explain how you can decide whether or not the players and the club have an equal chance of winning. Use a diagram if it helps clarify your explanation.

9. Challenging problem (problem solving, significant mathematical content)

   For any triangle there exists a point x such that the sum of the distances from each vertex to x is a minimum. Consider all isosceles triangles of perimeter 6 units. For which of these triangles is the sum of the distances from each vertex to x a minimum?

10. Investigations (multistage, open-ended)

For reasons of space, no example is provided for this task type. (Examples can be found in California Assessment Program, *A Sampler of Mathematics Assessment,* 1991.)

Investigations can be divided into three phases.

Phase 1 provides an opportunity for teacher and students to explore, in a series of guided familiarization activities, both context and any associated materials or equipment. Appropriate social organization and particularly suitable grouping structures may be negotiated at this time.

Phase 2 increases the sophistication of the tasks and provides both a guided investigation of the content and a clarification of the basic ideas. This phase concludes with a culminating question, which is open-ended but content specific. Students should be encouraged to evaluate their work in terms of the statement "I think I understand the content."

Phase 3 provides the opportunity for significant student construction and typically involves an open-ended investigation with some design component. Students should apply to their own work the criterion "I think I have successfully completed this task."

11. Fermi problems (context-specific, minimal cuing of mathematical tool skills)

How many piano tuners are there in Adelaide?

It is the intention with Fermi problems that students should not have recourse to other sources of information beyond the knowledge of the group with which they are working.

For a sample of a student response to this task, see the Fermi Problem box on the next page.

For the purpose of recording their problem-solving attempts, the students were asked to employ a five-part report format.

a. State the problem.
b. What do you know that would help you solve the problem?

Fermi Problem: Student Work Sample

Group Members: Casey, John, Steve, Susan

Fermi Problem 6: How Many Piano Tuners Are There in Adelaide?

What We Know Already

- Casey's piano is tuned once a year.
- The population of Adelaide is about 1 million.
- Our population is about 250,000.
- A piano tuner could tune 2 or 3 pianos a day (say 12 a week).
- An average family is about 5 people.

What We Did

1. We wrote down all the things we know that might help us solve the problem (Casey did this).
2. We worked out that there are probably 200,000 families in Adelaide.
3. We guessed that about $\frac{1}{15}$ of them have a piano.
4. So there would be 13,333 pianos in Adelaide (Susan did this).
5. We thought a piano tuner would work 48 weeks in a year. That's 576 pianos (John did this).
6. Steve worked out (on his calculator) that you would need 23.15 piano tuners.
7. We didn't know how many school or concert halls have pianos, but we thought 1 extra piano tuner would make up for these.

What We Found

We decided that there are probably 24 piano tuners in Adelaide.

How Good Is Our Answer?

We could be wrong about how many families have pianos. This is probably our biggest approximation. We had trouble checking our answer. If we knew about how many piano tuners were here, we could multiply by 4 to check our answer.

If we knew how much it cost to get a piano tuned, we could check to see whether a piano tuner would earn enough.

But we didn't know either of these. We think the number of piano tuners will definitely be between 20 and 30.

c. Record what you did, step by step.
d. State your answer.
e. How good is your answer?

With regard to this last requirement, students were offered the analogy "a chain is only as strong as its weakest link" and were asked to identify the biggest approximation they found necessary in generating their answer.

One of the characteristics that distinguishes Fermi problems is their capacity to reveal the mathematical tool skills a student spontaneously chooses to access. This characteristic is shared by such problems as "Which is the better fit: A square peg in a round hole or a round peg in a square hole?" (Schoenfeld 1985). As with the Good Question of example 5, these tasks are characterized by simple expression, leaving the responsibility for the elaboration of the task demands in the hands of the student.

12. Investigative project (open-ended, exploratory)
To complete an investigative project, students are provided with a theme, some general advice, and some possible starting points.

Theme: Is near enough good enough? Your project must be based on or incorporate the study of errors and approximations. It should use mathematics appropriate to the focus of the Change and Approximations blocks. You are encouraged to show initiative and be independent in carrying out your project.

General advice: Errors arise in different ways, and some are more significant than others. You should therefore distinguish between errors in data (due to human error and limitations in measuring equipment), errors created by calculations with incorrect data, and methods that are inherently approximate, such as truncation of series and replacement of curves by polygons.

The general criteria for assessment should be borne in mind, but here are some of the issues that occur specifically when considering errors and approximations.

- What is the effect of the error or approximation on the data and/or results?
- Are the errors acceptable?

Acknowledge which area of mathematics you use: coordinate geometry, calculus, or algebra. You may choose to develop a computer program to assist you, or you may use a recognized computer package, but remember to include your own analysis of the problem.

Starting points: You may investigate any topic related to the theme. You must discuss with your teacher your choice of topic and how it relates to the theme. The examples below show some starting points for projects. It is not compulsory to use the starting points.

A range of starting points are then listed under various headings. Here's an example.

Errors arising from inaccuracies of data

Some inaccuracies cause larger errors than others. How, for example, does the effect of a 1% error in setting a course for a ship compare with a 1% error in estimating the speed of the ocean?

- Compare the error of angle with the error in speed for a tennis serve.
- Compare the effect on predicted population due to errors in estimating birth rates, death rates, and so on.

The investigative project report format is specified in some detail and includes a section requiring the student to identify the mathematical methods used.

In choosing a task type for the purpose of assessment, it is useful to subject the task to the scrutiny of these four questions:

a. What aspects of mathematical performance are being assessed by this task type?

b. What elements of this task type are essential to the purpose?
c. What elements of this task type are optional for the purpose?
d. What elements are missing from this task type that might contribute to the purpose?

The preceding twelve task types appear to offer the opportunity for students to display a suitable range of mathematical performances. Among these tasks, examples 5, 8, 10, and 12 are categorized as open-ended for the purposes of this discussion. Any such categorization is open to challenge, and the identification of the characteristics of open-ended tasks with these examples is intended to be illustrative rather than definitive.